A. N. Whitehead was born on 15 February 1861, in Ramsgate, his father being Canon Whitehead, vicar of St Peter's, Isle of Thanet. He was educated at Sherborne School and Trinity College, Cambridge.

From 1885 until 1911 he was senior lecturer in mathematics at Trinity College. From 1911 until 1914 he was lecturer in applied mathematics and mechanics, later reader in geometry at University College, London. For ten years he was professor of applied mathematics at the Imperial College of Science. In 1924 he went to Harvard as professor of Philosophy. He retired in 1937, and died in 1947.

His works include *Principia Mathematica* (with Bertrand Russell), *The Principles of Natural Knowledge*, *Religion in the Making*, *Nature and Life*.

SCIENCE AND THE MODERN WORLD

Lowell Lectures, 1925

ALFRED NORTH WHITEHEAD

Collins
FONTANA BOOKS

Science and the Modern World was published by
the Syndics of the Cambridge University Press in 1926

First issued in Fontana Books 1975
Reprinted by permission of Cambridge University Press

Made and printed in Great Britain by
William Collins Sons & Co Ltd Glasgow

CONTENTS

TO

MY COLLEAGUES,

PAST AND PRESENT

WHOSE FRIENDSHIP IS INSPIRATION

PREFACE

The special value of Whitehead's philosophy comes from the fact that he was an excellent logician who was not afraid to be also an adventurous metaphysician. His philosophy of organism tries to provide that view of the meaning and significances of life which the layman anxiously looks for but most professionals fail to provide.

As a thinker and logician he could hardly have better credentials. With Bertrand Russell in the early years of this century he produced a gigantic work, *Principia Mathematica*, which tried to show how all mathematics can be derived from logic. It seems curious to us today that at that time and for many years longer these two great men, with almost or quite the most powerful brains in the world, proceeded on the assumption that language was transparent, as Russell later put it. You could use it like glass to find what was beyond. It did not obstruct the view.

Whitehead himself played a considerable part in showing us some of the barriers that language actually presents. In *Science and the Modern World* he introduces what he calls the fallacy of misplaced concreteness (p. 68). Once we have given a name to some abstract entity or quality we tend to apply to it the same grammatical treatment as we should give to an object that could be seen and touched. Think of all the troubles that have arisen from treating the words 'mind' and 'soul' in this way. And Whitehead proceeds to show us how the meanings given by science to 'matter' and 'space' have determined much of our intellectual and even social life. Appreciation of the caution that is necessary to avoid such mistakes led on to the school of philosophy labelled logical positivism, which flourished between 1930 through recent centuries. And a really brilliant account it is. To top all of his abilities Whitehead possessed an immense store of learning about the history of science, mathematics and philosophy. You will learn more in these 200-odd pages about the growth of science than from much larger historical works. But Whitehead was trying to do more than write a history. In the guise of a 'critic of cosmologies' he was trying to present a cosmology. Or perhaps if not quite so ambitious an aim, at least to show the basis upon which one might be built.

The basis was to look behind the assumptions of the seventeenth to nineteenth centuries that in nature 'there is merely motion of material.' (p. 71) And 'the enormous success of the scientific abstractions, yielding on the one hand matter with its simple location in space and time, on the other hand mind . . . has foisted on to philosophy the task of accepting them as the most concrete rendering of fact. Thereby, modern philosophy has been ruined.' (p.73) All this because of 'the ascription of misplaced concreteness to the scientific scheme of the seventeenth century'.

Whitehead's own answer was to develop what may be called a philosophy or organism deriving from Francis Bacon, Berkeley and Spinoza. It is not altogether easy for me to follow him here and the reader must judge for himself. He seeks for a real concreteness to avoid one that is 'misplaced'. If you 'start from the immediate facts of our psychological experience, as surely an empiricist should begin, you are at once led to the organic conception of nature' (p. 93). And 'the emergence of organisms depends on a selective activity which is akin to purpose.' (p. 133)

In the later chapters he deals with relativity and the quantum theory and seems to find through them some way of relating psychology to physiology and so to physics. It is true that the very difficulties in the way of using the concepts of modern atomic physics (proton, electron, neutron, quark, etc.) help us to avoid the danger of treating them as concrete. Moreover these entities (or concepts, and 1960. Ryle's treatment of 'the ghost in the machine' is a study in how to avoid misplaced concreteness.

But the special interest of Whitehead is that he used his immense brain power not only to solve the logical puzzles that sometimes seemed to obsess his colleague Russell, but to penetrate what we may call the meaning of life. His approach to this was through the work of Henri Bergson, the French philosopher-biologist. His *Creative Evolution* (1907) produced a philosophy of time and organism which has never really caught on among biologists. His solution was to postulate a vital principle, and this did not appeal when scientists were at last beginning to achieve understanding of the material forces at work in living things. But the real point, which Whitehead took up from Bergson, is that living systems have an organization that is inherited from their past and is comprehensible only by consideration of that past and hence of history and of time. As we should put it now, organization can only be maintained by information, which is the opposite of entropy, the tendency to disorder.

Whitehead was one of the first to emphasize that living systems have shown increasing order proceeding through time. He would have rejoiced to learn what we now know of the universal code of the genes and of speculations about how the order of life arose such as are discussed by Bernal in his *Origin of Life* (1967). Whitehead in his time believed that 'the aboriginal stuff, or material, from which a materialistic philosophy starts is incapable of evolution.' (p. 133) Indeed Bernal, like all of us, was still struggling with the problem of where the order of life comes from. But Whitehead was a physicist and philosopher, not a biologist and he tried to generalize from the concepts of order and organism in the living world to all natural phenomena. 'Philosophy' (he writes) 'in one of its functions, is the critic of cosmologies.' This gives the clue to the structure of the set of lectures which make up the book, and to its title. On the face of it, it is an account of the progress of science whatever they may be) are ordered together to make the world as we know it. In this sense we may begin to realize the presence of types of organism other than ourselves. And of course the cosmos shows us still wider order. Nevertheless Whitehead seems to be able to give to these ideas an application to all phenomena, which I confess not to find wholly satisfactory. 'In this alternative scheme, the notion of material, as fundamental, has been replaced by that of organic synthesis' (p. 189) and 'I am maintaining that the understanding of actuality requires a reference to ideality.' (p. 190) Yet he does not give specific examples of how these ideas are to be made to provide a satisfactory cosmology.

But the reader will have a wonderful time working out for himself just what is meant by this subtle thinker in his most famous book. At the very end he applies all his thinking to the specific problems of religion and finally to 'Requisites for Social Progress'. In this attention to human needs, as in so much else, Whitehead's thoughts are very applicable today. Indeed this is one of those books that never gets out of date. You will find on every page aphorisms and insights that will fill you with delight. His thoughts seem to move effortlessly along and yet the whole is a beautifully structured presentation of the drama of the human intellectual condition.

London, 1974 Professor J. Z. Young, M.A., F.R.S.

CHAPTER I

THE ORIGINS OF MODERN SCIENCE

The progress of civilization is not wholly a uniform drift towards better things. It may perhaps wear this aspect if we map it on a scale which is large enough. But such broad views obscure the details on which we rest our whole understanding of the process. New epochs emerge with comparative suddenness, if we have regard to the scores of thousands of years throughout which the complete history extends. Secluded races suddenly take their places in the main stream of events: technological discoveries transform the mechanism of human life: a primitive art quickly flowers into full satisfaction of same aesthetic craving: great religions in their crusading youth spread through the nations the peace of Heaven and the sword of the Lord.

The sixteenth century of our era saw the disruption of Western Christianity and the rise of modern science. It was an age of ferment. Nothing was settled, though much was opened – new worlds and new ideas. In science, Copernicus and Vesalius may be chosen as representative figures: they typify the new cosmology and the scientific emphasis on direct observation. Giordano Bruno was the martyr; though the cause for which he suffered was not that of science, but that of free imaginative speculation. His death in the year 1600 ushered in the first century of modern science in the strict sense of the term. In his execution there was an unconscious symbolism: for the subsequent tone of scientific thought has contained distrust of his type of general speculativeness. The Reformation, for all its importance, may be considered as a domestic affair of the European races. Even the Christianity of the East viewed it with profound disengagement. Furthermore, such disruptions are no new phenomena in the history of Christianity or of other religions. When we project this great revolution upon the whole history of the Christian Church, we cannot look upon it as introducing a new principle into human life. For good or for evil, it was a great transformation of religion; but it was not the coming of religion. It did not itself claim to be so. Reformers maintained that they were only restoring what had been forgotten.

It is quite otherwise with the rise of modern science. In every way it contrasts with the contemporary religious movement. The Reformation was a popular uprising, and for a century and a half drenched Europe in blood. The beginnings of the scientific movement were confined to a minority among the intellectual élite. In a generation which saw the Thirty Years' War and remembered Alva in the Netherlands, the worst that happened to men of science was that Galileo suffered an honourable detention and a mild reproof, before dying peacefully in his bed. The way in which the persecution of Galileo has been remembered is a tribute to the quiet commencement of the most intimate change in outlook which the human race had yet encountered. Since a babe was born in a manger, it may be doubted whether so great a thing has happened with so little stir.

The thesis which these lectures will illustrate is that this quiet growth of science has practically recoloured our mentality so that modes of thought which in former times were exceptional, are now broadly spread through the educated world. This new colouring of ways of thought had been proceeding slowly for many ages in the European peoples. At last it issued in the rapid development of science; and has thereby strengthened itself by its most obvious application. The new mentality is more important even than the new science and the new technology. It has altered the metaphysical presuppositions and the imaginative contents of our minds; so that now the old stimuli provoke a new response. Perhaps my metaphor of a new colour is too strong. What I mean is just that slightest change of tone which yet makes all the difference. This is exactly illustrated by a sentence from a published letter of that adorable genius, William James. When he was finishing his great treatise on the *Principles of Psychology*, he wrote to his brother Henry James, 'I have to forge every sentence in the teeth of irreducible and stubborn facts.'

This new tinge to modern minds is a vehement and passionate interest in the relation of general principles to irreducible and stubborn facts. All the world over and at all times there have been practical men, absorbed in 'irreducible and stubborn facts'; all the world over and at all times there have been men of philosophic temperament who have been absorbed in the weaving of general principles. It is this union of passionate interest in the detailed facts with equal devotion to abstract generalization which forms the novelty in our present society. Previously it had appeared sporadically and as if by chance. This balance of mind has now become part of the tradition which infects cultivated thought. It is the salt which keeps life sweet. The main business of universities is to transmit this tradition as a widespread inheritance from generation to generation.

Another contrast which singles out science from among the European movements of the sixteenth and seventeenth centuries, is its unversality. Modern science was born in Europe, but its home is the whole world. In the last two centuries there has been a long and confused impact of Western modes upon the civilization of Asia. The wise men of the East have been puzzling, and are puzzling, as to what may be the regulative secret of life which can be passed from West to East without the wanton destruction of their own inheritance which they so rightly prize. More and more it is becoming evident that what the West can most readily give to the East is its science and its scientific outlook. This is transferable from country to country, and from race to race, wherever there is a rational society.

In this course of lectures I shall not discuss the details

of scientific discovery. My theme is the energizing of a state of mind in the modern world, its broad generalizations, and its impact upon other spiritual forces. There are two ways of reading history, forwards and backwards. In the history of thought, we require both methods. A climate of opinion – to use the happy phrase of a seventeenth-century writer – requires for its understanding the consideration of its antecedents and its issues. Accordingly in this lecture I shall consider some of the antecedents of our modern approach to the investigation of nature.

In the first place, there can be no living science unless there is a widespread instinctive conviction in the existence of an order of things, and, in particular, of an order of nature. I have used the word *instinctive* advisedly. It does not matter what men say in words, so long as their activities are controlled by settled instincts. The words may ultimately destroy the instincts. But until this has occurred, words do not count. This remark is important in respect to the history of scientific thought. For we shall find that since the time of Hume, the fashionable scientific philosophy has been such as to deny the rationality of science. This conclusion lies upon the surface of Hume's philosophy. Take, for example, the following passage from Section IV of his *Inquiry Concerning Human Understanding*:

> In a word, then, every effect is a distinct event from its cause. It could not, therefore, be discovered in the cause; and the first invention or conception of it, *a priori*, must be entirely arbitrary.

If the cause in itself discloses no information as to the effect, so that the first invention of it must be *entirely* arbitrary, it follows at once that science is impossible, except in the sense of establishing *entirely arbitrary* connections which are not warranted by anything intrinsic to the natures either of causes or effects. Some variant of Hume's philosophy has generally prevailed among men of science. But scientific faith has risen to the occasion, and has tacitly removed the philosophic mountain.

In view of this strange contradiction in scientific thought, it is of the first importance to consider the antecedents of a faith which is impervious to the demand for a consistent rationality. We have therefore to trace the rise of the instinctive faith that there is an order of nature which can be traced in every detailed occurrence.

Of course we all share in this faith, and we therefore believe that the reason for the faith is our apprehension of its truth. But the formation of a general idea – such as the idea of the order of nature – and the grasp of its importance, and the observation of its exemplification in a variety of occasions are by no means the necessary consequences of the truth of the idea in question. Familiar things happen, and mankind does not bother about them. It requires a very unusual mind to undertake the analysis of the obvious. Accordingly I wish to consider the stages in which this analysis became explicit, and finally became unalterably impressed upon the educated minds of Western Europe.

Obviously, the main recurrences of life are too insistent to escape the notice of the least rational of humans; and even before the dawn of rationality, they have impressed themselves upon the instincts of animals. It is unnecessary to labour the point, that in broad outline certain general states of nature recur, and that our very natures have adapted themselves to such repetitions.

But there is a complementary fact which is equally true and equally obvious: nothing ever really recurs in exact detail. No two days are identical, no two winters. What has gone, has gone for ever. Accordingly the practical philosophy of mankind has been to expect the broad recurrences, and to accept the details as emanating from the inscrutable womb of things behind the ken of rationality. Men expected the sun to rise, but the wind bloweth where it listeth.

Certainly from the classical Greek civilization onwards

there have been men, and indeed groups of men, who have placed themselves beyond this acceptance of an ultimate irrationality. Such men have endeavoured to explain all phenomena as the outcome of an order of things which extends to every detail. Geniuses such as Aristotle, or Archimedes, or Roger Bacon, must have been endowed with the full scientific mentality, which instinctively holds that all things great and small are conceivable as exemplifications of general principles which reign throughout the natural order.

But until the close of the Middle Ages the general educated public did not feel that intimate conviction, and that detailed interest, in such an idea, so as to lead to an unceasing supply of men, with ability and opportunity adequate to maintain a co-ordinated search for the discovery of these hypothetical principles. Either people were doubtful about the existence of such principles, or were doubtful about any success in finding them, or took no interest in thinking about them, or were oblivious to their practical importance when found. For whatever reason, search was languid, if we have regard to the opportunities of a high civilization and the length of time concerned. Why did the pace suddenly quicken in the sixteenth and seventeenth centuries? At the close of the Middle Ages a new mentality discloses itself. Invention stimulated thought, thought quickened physical speculation, Greek manuscripts disclosed what the ancients had discovered. Finally although in the year 1500 Europe knew less than Archimedes who died in the year 212 BC, yet in the year 1700, Newton's *Principia* had been written and the world was well started on the modern epoch.

There have been great civilizations in which the peculiar balance of mind required for science has only fitfully appeared and has produced the feeblest result. For example, the more we know of Chinese art, of Chinese literature, and of the Chinese philosophy of life, the more we admire the heights to which that civilization attained. For thousands of years, there have been in China acute and learned men patiently devoting their lives to study. Having regard to the span of time, and to the population concerned, China forms the largest volume of civilization which the world has seen. There is no reason to doubt the intrinsic capacity of individual Chinese for the pursuit of science. And yet Chinese science is practically negligible. There is no reason to believe that China if left to itself would have ever produced any progress in science. The same may be said of India. Furthermore, if the Persians had enslaved the Greeks, there is no definite ground for belief that science would have flourished in Europe. The Romans showed no particular originality in that line. Even as it was, the Greeks, though they founded the movement, did not sustain it with the concentrated interest which modern Europe has shown. I am not alluding to the last few generations of the European peoples on both sides of the ocean; I mean the smaller Europe of the Reformation period, distracted as it was with wars and religious disputes. Consider the world of the Eastern Mediterranean, from Sicily to Western Asia, during the period of about 1400 years from the death of Archimedes [in 212 BC] to the irruption of the Tartars. There were wars and revolutions and large changes of religion : but nothing much worse than the wars of the sixteenth and seventeenth centuries throughout Europe. There was a great and wealthy civilization, pagan, Christian, Mahometan. In that period a great deal was added to science. But on the whole the progress was slow and wavering; and, except in mathematics, the men of the Renaissance practically started from the position which Archimedes had reached. There had been some progress in medicine and some progress in astronomy. But the total advance was very little compared to the marvellous success of the seventeenth century. For example, compare the progress of scientific knowledge from the year 1560, just before the births of Galileo and of Kepler, up to the year 1700, when Newton was in the height of his fame, with the progress in the ancient period, already mentioned, exactly ten times as long.

Nevertheless, Greece was the mother of Europe; and it is to Greece that we must look in order to find the origin of our modern ideas. We all know that on the eastern shores of the Mediterranean there was a very flourishing school of Ionian philosophers, deeply interested in theories concerning nature. Their ideas have been transmitted to us, enriched by the genius of Plato and Aristotle. But, with the exception of Aristotle, and it is a large exception, this school of thought had not attained to the complete scientific mentality. In some ways, it was better. The Greek genius was philosophical, lucid and logical. The men of this group were primarily asking philosophical questions. What is the substratum of nature? Is it fire, or earth, or water, or some combination of any two, or of all three? Or is it a mere flux, not reducible to some static material? Mathematics interested them mightily. They invented its generality, analysed its premises, and made notable discoveries of theorems by a rigid adherence to deductive reasoning. Their minds were infected with an eager generality. They demanded clear, bold ideas, and strict reasoning from them. All this was excellent; it was genius; it was ideal preparatory work. But it was not science as we understand it. The patience of minute observation was not nearly so prominent. Their genius was not so apt for the state of imaginative muddled suspense which precedes successful inductive generalization. They were lucid thinkers and bold reasoners.

Of course there were exceptions, and at the very top: for example, Aristotle and Archimedes. Also for patient observation, there were the astronomers. There was a mathematical lucidity about the stars, and a fascination about the small numerable band of runaway planets.

Every philosophy is tinged with the colouring of some secret imaginative background, which never emerges explicitly into its trains of reasoning. The Greek view of nature, at least that cosmology transmitted from them to later ages, was essentially dramatic. It is not necessarily wrong for this reason: but it was overwhelmingly dramatic. It thus conceived nature as articulated in the way of a work of dramatic art, for the exemplification of general ideas converging to an end. Nature was differentiated so as to provide its proper end for each thing. There was the centre of the universe as the end of motion for those things which are heavy, and the celestial spheres as the end of motion for those things whose natures lead them upwards. The celestial spheres were for things which are impassible and ingenerable, the lower regions for things passible and generable. Nature was a drama in which each thing played its part.

I do not say that this is a view to which Aristotle would have subscribed without severe reservations, in fact without the sort of reservations which we ourselves would make. But it was the view which subsequent Greek thought extracted from Aristotle and passed on to the Middle Ages. The effect of such an imaginative setting for nature was to damp down the historical spirit. For it was the end which seemed illuminating, so why bother about the beginning? The Reformation and the scientific movement were two aspects of the historical revolt which was the dominant intellectual movement of the later Renaissance. The appeal to the origins of Christianity, and Francis Bacon's appeal to efficient causes as against final causes, were two sides of one movement of thought. Also for this reason Galileo and his adversaries were at hopeless cross purposes, as can be seen from his *Dialogues on the Two Systems of the World*.

Galileo keeps harping on how things happen, whereas his adversaries had a complete theory as to why things happen. Unfortunately the two theories did not bring out the same results. Galileo insists upon 'irreducible and stubborn facts,' and Simplicius, his opponent, brings forward reasons, completely satisfactory, at least to himself. It is a great mistake to conceive this historical revolt as an appeal to reason. On the contrary, it was through and through an anti-intellectualist movement. It was the return to the contemplation of brute fact; and it was based on a recoil from the inflexible rationality of medieval thought. In making this statement I am merely summarizing what at the time the adherents of the old régime themselves asserted. For example, in the fourth book of Father Paul Sarpi's *History of the Council of Trent*, you will find that in the year 1551 the Papal Legates who presided over the Council ordered:

> That the Divines ought to confirm their opinions with the holy Scripture, Traditions of the Apostles, sacred and approved Councils, and by the Constitutions and Authorities of the holy Fathers; that they ought to use brevity, and avoid superfluous and unprofitable questions, and perverse contentions . . . This order did not please the Italian Divines; who said it was a novity, and a condemning of School-Divinity, which, in all difficulties, *useth reason*, and because it was not lawful [*i.e.* by this decree] to treat as St Thomas [Aquinas], St Bonaventure, and other famous men did.

It is impossible not to feel sympathy with these Italian divines, maintaining the lost cause of unbridled rationalism. They were deserted on all hands. The Protestants were in full revolt against them. The Papacy failed to support them, and the Bishops of the Council could not even understand them. For a few sentences below the foregoing quotation, we read: 'Though many complained here-of [i.e. of the Decree], yet it prevailed but little, because generally the Fathers [i.e. the Bishops] desired to hear men speak with intelligible terms, not abstrusely, as in the matter of Justification, and others already handled.'

Poor belated medievalists! When they used reason they were not even intelligible to the ruling powers of their epoch. It will take centuries before stubborn facts are reducible by reason, and meanwhile the pendulum swings slowly and heavily to the extreme of the historical method.

Forty-three years after the Italian divines had written

this memorial, Richard Hooker in his famous *Laws of Ecclesiastical Polity* makes exactly the same complaint of his Puritan adversaries[1]. Hooker's balanced thought – from which the appellation 'The Judicious Hooker' is derived, and his diffuse style, which is the vehicle of such thought, make his writings singularly unfit for the process of summarizing by a short, pointed quotation. But, in the section referred to, he reproaches his opponents with *Their Disparagement of Reason*; and in support of his own position, definitely refers to 'The greatest amongst the school-divines' by which designation I presume that he refers to St Thomas Aquinas.

Hooker's *Ecclesiastical Polity* was published just before Sarpi's *Council of Trent*. Accordingly there was complete independence between the two works. But both the Italian divines of 1551, and Hooker at the end of that century testify to the anti-rationalist trend of thought at that epoch, and in this respect contrast their own age with the epoch of scholasticism.

This reaction was undoubtedly a very necessary corrective to the unguarded rationalism of the Middle Ages. But reactions run to extremes. Accordingly, although one outcome of this reaction was the birth of modern science, yet we must remember that science thereby inherited the bias of thought to which it owes its origin.

The effect of Greek dramatic literature was many-sided so far as concerns the various ways in which it indirectly affected medieval thought. The pilgrim fathers of the scientific imagination as it exists today, are the great tragedians of ancient Athens, Aeschylus, Sophocles, Euripides. Their vision of fate, remorseless and indifferent, urging a tragic incident to its inevitable issue, is the vision possessed by science. Fate in Greek tragedy becomes the order of nature in modern thought. The absorbing interest in the particular heroic incidents, as an example and a verification of the workings of fate, reappears in our epoch as concentration

[1] Cf. Book III, Section viii.

of interest on the crucial experiments. It was my good fortune to be present at the meeting of the Royal Society in London when the Astronomer Royal for England announced that the photographic plates of the famous eclipse, as measured by his colleagues in Greenwich Observatory, had verified the prediction of Einstein that rays of light are bent as they pass in the neighbourhood of the sun. The whole atmosphere of tense interest was exactly that of the Greek drama : we were the chorus commenting on the decree of destiny as disclosed in the development of a supreme incident. There was dramatic quality in the very staging : the traditional ceremonial, and in the background the picture of Newton to remind us that the greatest of scientific generalizations was now, after more than two centuries, to receive its first modification. Nor was the personal interest wanting : a great adventure in thought had at length come safe to shore.

Let me here remind you that the essence of dramatic tragedy is not unhappiness. It resides in the solemnity of the remorseless working of things. This inevitableness of destiny can only be illustrated in terms of human life by incidents which in fact involve unhappiness. For it is only by them that the futility of escape can be made evident in the drama. This remorseless inevitableness is what pervades scientific thought. The laws of physics are the decrees of fate.

The conception of the moral order in the Greek plays was certainly not a discovery of the dramatists. It must have passed into the literary tradition from the general serious opinion of the times. But in finding this magnificent expression, it thereby deepened the stream of thought from which it arose. The spectacle of a moral order was impressed upon the imagination of classical civilization.

The time came when that great society decayed, and Europe passed into the Middle Ages. The direct influence of Greek literature vanished. But the concept of the moral order and of the order of nature had enshrined itself in the Stoic philosophy. For example, Lecky in his *History of European Morals* tells us 'Seneca maintains that the Divinity has determined all things by an inexorable law of destiny, which He has decreed, but which He Himself obeys.' But the most effective way in which the Stoics influenced the mentality of the Middle Ages was by the diffused sense of order which arose from Roman law. Again to quote Lecky, 'The Roman legislation was in a twofold manner the child of philosophy. It was in the first place formed upon the philosophical model, for, instead of being a mere empirical system adjusted to the existing requirements of society, it laid down abstract principles of right to which it endeavoured to conform; and, in the next place, these principles were borrowed directly from Stoicism.' In spite of the actual anarchy throughout large regions in Europe after the collapse of the Empire, the sense of legal order always haunted the racial memories of the imperial populations. Also the Western Church was always there as a living embodiment of the traditions of Imperial rule.

It is important to notice that this legal impress upon medieval civilization was not in the form of a few wise precepts which should permeate conduct. It was the conception of a definite articulated system which defines the legality of the detailed structure of social organism, and of the detailed way in which it should function. There was nothing vague. It was not a question of admirable maxims, but of definite procedure to put things right and to keep them there. The Middle Ages formed one long training of the intellect of Western Europe in the sense of order. There may have been some deficiency in respect to practice. But the idea never for a moment lost its grip. It was preeminently an epoch of orderly thought, rationalist through and through. The very anarchy quickened the sense for coherent system; just as the modern anarchy of Europe has stimulated the intellectual vision of a League of Nations.

But for science something more is wanted than a general sense of the order in things. It needs but a sentence to point out how the habit of definite exact thought was implanted in the European mind by the long dominance of scholastic logic and scholastic divinity. The habit remained after the philosophy had been repudiated, the priceless habit of looking for an exact point and of sticking to it when found. Galileo owes more to Aristotle than appears on the surface of his *Dialogues*: he owes to him his clear head and his analytic mind.

I do not think, however, that I have even yet brought out the greatest contribution of medievalism to the formation of the scientific movement. I mean the inexpugnable belief that every detailed occurrence can be correlated with its antecedents in a perfectly definite manner, exemplifying general principles. Without this belief the incredible labours of scientists would be without hope. It is this instinctive conviction, vividly poised before the imagination, which is the motive power of research: that there is a secret, a secret which can be unveiled. How has this conviction been so vividly implanted in the European mind?

When we compare this tone of thought in Europe with the attitude of other civilizations when left to themselves, there seems but one source for its origin. It must come from the medieval insistence on the rationality of God, conceived as with the personal energy of Jehovah and with the rationality of a Greek philosopher. Every detail was supervised and ordered: the search into nature could only result in the vindication of the faith in rationality. Remember that I am not talking of the explicit beliefs of a few individuals. What I mean is the impress on the European mind arising from the unquestioned faith of centuries. By this I mean the instinctive tone of thought and not a mere creed of words.

In Asia, the conceptions of God were of a being who was either too arbitrary or too impersonal for such ideas to have much effect on instinctive habits of mind. Any definite occurrence might be due to the fiat of an irrational despot, or might issue from some impersonal, inscrutable origin of things. There was not the same confidence as in the intelligible rationality of a personal being. I am not arguing that the European trust in the scrutability of nature was logically justified even by its own theology. My only point is to understand how it arose. My explanation is that the faith in the possibility of science, generated antecedently to the development of modern scientific theory, is an unconscious derivative from medieval theology.

But science is not merely the outcome of instinctive faith. It also requires an active interest in the simple occurrences of life for their own sake.

This qualification 'for their own sake' is important. The first phase of the Middle Ages was an age of symbolism. It was an age of vast ideas, and of primitive technique. There was little to be done with nature, except to coin a hard living from it. But there were realms of thought to be explored, realms of philosophy and realms of theology. Primitive art could symbolize those ideas which filled all thoughtful minds. The first phase of medieval art has a haunting charm beyond compare : its own intrinsic quality is enhanced by the fact that its message, which stretched beyond art's own self-justification of aesthetic achievement, was the symbolism of things lying behind nature itself. In this symbolic phase, medieval art energized in nature as its medium, but pointed to another world.

In order to understand the contrast between these early Middle Ages and the atmosphere required by the scientific mentality, we should compare the sixth century in Italy with the sixteenth century. In both centuries the Italian genius was laying the foundations of a new epoch. The history of the three centuries preceding the earlier period, despite the promise for the future introduced by the rise of Christianity, is overwhelmingly infected by the sense of the decline of civilization. In each generation something has been lost. As we read the records, we are haunted by the shadow of the coming barbarism. There are great men, with fine achievements in action or in thought. But their total effect is merely for some short time to arrest the general decline. In the sixth century we are, so far as Italy is concerned, at the lowest point of the curve. But in that century every action is laying the foundation for the tremendous rise of the new European civilization. In the background the Byzantine Empire, under Justinian, in three ways determined the character of the early Middle Ages in Western Europe. In the first place, its armies, under Belisarius and Narses, cleared Italy from the Gothic domination. In this way, the stage was freed for the exercise of the old Italian genius for creating organizations which shall be protective of ideals of cultural activity. It is impossible not to sympathize with the Goths: yet there can be no doubt but that a thousand years of the Papacy were infinitely more valuable for Europe than any effects derivable from a well-established Gothic kingdom of Italy.

In the second place, the codification of the Roman law established the ideal of legality which dominated the sociological thought of Europe in the succeeding centuries. Law is both an engine for government, and a condition restraining government. The canon law of the Church, and the civil law of the State, owe to Justinian's lawyers their influence on the development of Europe. They established in the Western mind the ideal that an authority should be at once lawful, and law-enforcing, and should in itself exhibit a rationally adjusted system of organization. The sixth century in Italy gave the initial exhibition of the way in which the impress of these ideas was fostered by contact with the Byzantine Empire.

Thirdly, in the non-political spheres of art and learning Constantinople exhibited a standard of realized achievement which, partly by the impulse to direct imitation, and partly by the indirect inspiration arising from the mere knowledge that such things existed, acted as a perpetual spur to Western culture. The wisdom of the Byzantines, as it stood in the imagination of the first phase of medieval mentality, and the wisdom of the Egyptians as it stood in the imagination of the early Greeks, played analogous roles. Probably the actual knowledge of these respective wisdoms was, in either case, about as much as was good for the recipients. They knew enough to know the sort of standards which are attainable, and not enough to be fettered by static and traditional ways of thought. Accordingly, in both cases men went ahead on their own and did better. No account of the rise of the European scientific mentality can omit some notice of this influence of the Byzantine civilization in the background. In the sixth century there is a crisis in the history of the relations between the Byzantines and the West; and this crisis is to be contrasted with the influence of Greek literature on European thought in the fifteenth and sixteenth centuries. The two outstanding men, who in the Italy of the sixth century laid the foundations of the future, were St Benedict and Gregory the Great. By reference to them, we can at once see how absolutely in ruins was the approach to the scientific mentality which had been attained by the Greeks. We are at the zero point of scientific temperature. But the life-work of Gregory and of Benedict contributed elements to the reconstruction of Europe which secured that this reconstruction, when it arrived, should include a more effective scientific mentality than that of the ancient world. The Greeks were over-theoretical. For them science was an offshoot of philosophy. Gregory and Benedict were practical men, with an eye for the importance of ordinary things; and they combined this practical temperament with their religious and cultural activities. In particular, we owe it to St Benedict that the monasteries were the homes of practical agriculturalists, as well as of saints and of artists and men of learning. The alliance of science with technology, by which learning is kept in contact with irreducible and stubborn facts, owes much to the practical bent of the early Benedictines. Modern science derives from Rome as well as from Greece, and this Roman strain explains its gain in an energy of thought kept closely in contact with the world of facts.

But the influence of this contact between the monasteries and the facts of nature showed itself first in art. The rise of naturalism in the later Middle Ages was the entry into the European mind of the final ingredient necessary for the rise of science. It was the rise of interest in natural objects and in natural occurrences, for their own sakes. The natural foliage of a district was sculptured in out-of-the-way spots of the later buildings, merely as exhibiting delight in those familiar objects. The whole atmosphere of every art exhibited a direct joy in the apprehension of the things which lie around us. The craftsmen who executed the late medieval decorative sculpture, Giotto, Chaucer, Wordsworth, Walt Whitman, and, at the present day, the New England poet Robert Frost, are all akin to each other in this respect. The simple immediate facts are the topics of interest, and these reappear in the thought of science as the 'irreducible stubborn facts'.

The mind of Europe was now prepared for its new venture of thought. It is unnecessary to tell in detail the various incidents which marked the rise of science: the growth of wealth and leisure; the expansion of universities; the invention of printing; the taking of Constantinople; Copernicus; Vasco da Gama; Columbus; the telescope. The soil, the climate, the seeds were there, and the forest grew. Science has never shaken off the impress of its origin in the historical revolt of the later Renaissance. It has remained predominantly an anti-rationalistic movement, based upon a naïve faith. What reasoning it has wanted, has been borrowed from mathematics which is a surviving relic of Greek rationalism, following the deductive method. Science repudiates philosophy. In other words, it has never cared to justify its faith or to explain its meanings: and has remained blandly indifferent to its refutation by Hume.

Of course the historical revolt was fully justified. It was wanted. It was more than wanted: it was an absolute necessity for healthy progress. The world required centuries of contemplation of irreducible and stubborn facts. It is difficult for men to do more than one thing at a time, and that was the sort of thing they had to do after the rationalistic orgy of the Middle Ages. It was a very sensible reaction; but it was not a protest on behalf of reason.

There is, however, a Nemesis which waits upon those who deliberately avoid avenues of knowledge. Oliver Cromwell's cry echoes down the ages, 'My brethren, by the bowels of Christ I beseech you, bethink you that you may be mistaken.'

The progress of science has now reached a turning point. The stable foundations of physics have broken up: also for the first time physiology is asserting itself as an effective body of knowledge, as distinct from a scrap-heap. The old foundations of scientific thought are becoming unintelligible. Time, space, matter, material, ether, electricity, mechanism, organism, configuration, structure, pattern, function, all require reinterpretation. What is the sense of talking about a mechanical explanation when you do not know what you mean by mechanics?

The truth is that science started its modern career by taking over ideas derived from the weakest side of the philosophies of Aristotle's successors. In some respects it was a happy choice. It enabled the knowledge of the seventeenth century to be formularized so far as physics and chemistry were concerned, with a completeness which has lasted to the present time. But the progress of biology and psychology has probably been checked by the uncritical assumption of half-truths. If science is not to degenerate into a medley of *ad hoc* hypotheses, it must become philosophical and must enter upon a thorough criticism of its own foundations.

In the succeeding lectures of this course, I shall trace the successes and the failures of the particular conceptions of cosmology with which the European intellect has clothed itself in the last three centuries. General climates of opinion persist for periods of about two to three generations, that is to say, for periods of sixty to a hundred years. There are also shorter waves of thought, which play on the surface of the tidal movement. We shall find, therefore, transformations in the European outlook, slowly modifying the successive centuries. There persists, however, throughout the whole period the fixed scientific cosmology which presupposes the ultimate fact of an irreducible brute matter, or material, spread throughout space in a flux of configurations. In itself such a material is senseless, valueless, purposeless. It just does what it does do, following a fixed routine imposed by external relations which do not spring from the nature of its being. It is this assumption that I call 'scientific materialism'. Also it is an assumption which I shall challenge as being entirely unsuited to the scientific situation at which we have now arrived. It is not wrong, if properly construed. If we confine ourselves to certain types of facts, abstracted from the complete circumstances in which they occur, the materialistic assumption expresses these facts to perfection. But when we pass beyond the abstraction, either by more subtle employment of our senses, or by the request for meanings and for coherence of thoughts, the scheme breaks down at once. The narrow efficiency of the scheme was the very cause of its supreme methodological success. For it directed attention to just those groups of facts which, in the state of knowledge then existing, required investigation.

The success of the scheme has adversely affected the various currents of European thought. The historical revolt was anti-rationalistic, because the rationalism of the scholastics required a sharp correction by contact with brute fact. But the revival of philosophy in the hands of Descartes and his successors was entirely coloured in its development by the acceptance of the scientific cosmology at its face value. The success of their ultimate ideas confirmed scientists in their refusal to modify them as the result of an inquiry into their rationality. Every philosophy was bound in some way or other to swallow them whole. Also the example of science affected other regions of thought. The historical revolt has thus been exaggerated into the exclusion of philosophy from its proper role of harmonizing the various abstractions of methodological thought. Thought is abstract; and the intolerant use of abstractions is the major vice of the intellect. This vice is not wholly corrected by the recurrence to concrete experience. For after all, you need only attend to those aspects of your concrete experience which lie within some limited scheme. There are two methods for the purification of ideas. One of them is dispassionate observation by means of the bodily senses. But observation is selection. Accordingly, it is difficult to transcend a scheme of abstraction whose success is sufficiently wide. The other method is by comparing the various schemes of abstraction which are well founded in our various types of experience. This comparison takes the form of satisfying the demands of the Italian scholastic divines whom Paul Sarpi mentioned. They asked that reason should be used. Faith in reason is the trust that the ultimate natures of things lie together in a harmony which excludes mere arbitrariness. It is the faith that at the base of things we shall not find mere arbitrary mystery. The faith in the order of nature which has made possible the growth of science is a particular example of a deeper faith. This faith cannot be justified by any inductive generalization. It springs from direct inspection of the nature of things as disclosed in our own immediate present experience. There is no parting from your own shadow. To experience this faith is to know that in being ourselves we are more than ourselves: to know that our experience, dim and fragmentary as it is, yet sounds the utmost depths of reality: to know that detached details merely in order to be themselves demand that they should find themselves in a system of things: to know that this system includes the harmony of logical rationality, and the harmony of aesthetic achievement: to know that, while the harmony of logic lies upon the universe as an iron necessity, the aesthetic harmony stands before it as a living ideal moulding the general flux in its broken progress towards finer, subtler issues.

CHAPTER II

MATHEMATICS AS AN ELEMENT IN THE HISTORY OF THOUGHT

The science of pure mathematics, in its modern developments, may claim to be the most original creation of the human spirit. Another claimant for this position is music. But we will put aside all rivals, and consider the ground on which such a claim can be made for mathematics. The originality of mathematics consists in the fact that in mathematical science connections between things are exhibited which, apart from the agency of human reason, are extremely unobvious. Thus the ideas, now in the minds of contemporary mathematicians, lie very remote from any notions which can be immediately derived by perception through the senses; unless indeed it be perception stimulated and guided by antecedent mathematical knowledge. This is the thesis which I proceed to exemplify.

Suppose we project our imagination backwards through many thousands of years, and endeavour to realize the simple-mindedness of even the greatest intellects in those early societies. Abstract ideas which to us are immediately obvious must have been, for them, matters only of the most dim apprehension. For example take the question of number. We think of the number 'five' as applying to appropriate groups of any entities whatsoever – to five fishes, five children, five apples, five days. Thus in considering the relations of the number 'five' to the number 'three', we are thinking of two groups of things, one with five members and the other with three members. But we are entirely abstracting from any consideration of any particular entities, or even of any particular sorts of entities, which go to make up the membership of either of the two groups. We are merely thinking of those relationships between those two groups which are entirely independent of the individual essences of any of the members of either group. This is a very remarkable feat of abstraction; and it must have taken ages for the human race to rise to it. During a long period, groups of fishes will have been compared to each other in respect to their multiplicity, and groups of days to each other. But the first man who noticed the analogy between a group of seven fishes and a group of seven days made a notable advance in the history of thought. He was the first man who entertained a concept belonging to the science of pure mathematics. At that moment it must have been impossible for him to divine the complexity and subtlety of these abstract mathematical ideas which were waiting for discovery. Nor could he have guessed that these notions would exert a widespread fascination in each succeeding generation. There is an erroneous literary tradition which represents the love of mathematics as a monomania confined to a few eccentrics in each generation. But be this as it may, it would have been impossible to anticipate the pleasure derivable from a type of abstract thinking which had no counterpart in the then existing society. Thirdly, the tremendous future effect of mathematical knowledge on the lives of men, on their daily avocations, on their habitual thoughts, on the organization of society, must have been even more completely shrouded from the foresight of those early thinkers. Even now there is a very wavering grasp of the true position of mathematics as an element in the history of thought. I will not go so far as to say that to construct a history of thought without profound study of the mathematical ideas of successive epochs is like omitting Hamlet from the play which is named after him. That would be claiming too much. But it is certainly analogous to cutting out the part of Ophelia. This simile is singularly exact. For Ophelia is quite essential to the play, she is very charming – and a little mad. Let us grant that the pursuit of mathematics is a divine madness of the human spirit, a refuge from the goading urgency of contingent happenings.

When we think of mathematics, we have in our mind a science devoted to the exploration of number, quantity, geometry, and in modern times also including investigation into yet more abstract concepts of order, and into analogous types of purely logical relations. The point of mathematics is that in it we have always got rid of the particular instance, and even of any particular sorts of entities. So that for example, no mathematical truths apply merely to fish, or merely to stones, or merely to colours. So long as you are dealing with pure mathematics, you are in the realm of complete and absolute abstraction. All you assert is, that reason insists on the admission that, if any entities whatever have any relations which satisfy such-and-such purely abstract conditions, then they must have other relations which satisfy other purely abstract conditions.

Mathematics is thought moving in the sphere of complete abstraction from any particular instance of what it is talking about. So far is this view of mathematics from being obvious, that we can easily assure ourselves that it is not, even now, generally understood. For example, it is habitually thought that the certainty of mathematics is a reason for the certainty of our geometrical knowledge of the space of the physical universe. This is a delusion which has vitiated much philosophy in the past, and some philosophy in the present. This question of geometry is a test case of some urgency. There are certain alternative sets of purely abstract conditions possible for the relationship of groups of unspecified entities, which I will call geometrical conditions. I give them this name because of their general analogy to those conditions, which we believe to hold respecting the particular geometrical relations of things observed by us in our direct perception of nature. So far as our observations are concerned, we are not quite accurate enough to be certain of the exact conditions regulating the things we come across in nature. But we can by a slight stretch of hypothesis identify these observed conditions with some one set of the purely abstract geometrical conditions. In doing so, we make a particular determination of the group of unspecified entities which are the *relata* in the abstract science. In the pure mathematics of geometrical relationships, we say that, if *any* group entities enjoy *any* relationships among its members satisfying *this* set of abstract geometrical conditions, then such-and-such additional abstract conditions must also hold for such relationships. But when we come to physical space, we say that some definitely observed group of physical entities enjoys some definitely observed relationships among its members which do satisfy this above-mentioned set of abstract geometrical conditions. We thence conclude that the additional relationships which we concluded to hold in *any* such case, must therefore hold in *this particular* case.

The certainty of mathematics depends upon its complete abstract generality. But we can have no *a priori* certainty that we are right in believing that the observed entities in the concrete universe form a particular instance of what falls under our general reasoning. To take another example from arithmetic. It is a general abstract truth of pure mathematics that any group of forty entities can be subdivided into two groups of twenty entities. We are therefore justified in concluding that a particular group of apples which we believe to contain forty members can be subdivided into two groups of apples of which each contains twenty members. But there always remains the possibility that we have miscounted the big group; so that, when we come in practice to subdivide it, we shall find that one of the two heaps has an apple too few or an apple too many.

Accordingly, in criticizing an argument based upon the application of mathematics to particular matters of fact there are always three processes to be kept perfectly distinct in our minds. We must first scan the purely mathematical reasoning to make sure that there are no mere slips in it –

no casual illogicalities due to mental failure. Any mathematician knows from bitter experience that, in first elaborating a train of reasoning, it is very easy to commit a slight error which yet makes all the difference. But when a piece of mathematics has been revised, and has been before the expert world for some time, the chance of a casual error is almost negligible. The next process is to make quite certain of all the abstract conditions which have been presupposed to hold. This is the determination of the abstract premises from which the mathematical reasoning proceeds. This is a matter of considerable difficulty. In the past quite remarkable oversights have been made, and have been accepted by generations of the greatest mathematicians. The chief danger is that of oversight, namely, tacitly to introduce some condition, which it is natural for us to presuppose, but which in fact need not always be holding. There is another opposite oversight in this connection which does not lead to error, but only to lack of simplification. It is very easy to think that more postulated conditions are required than is in fact the case. In other words, we may think that some abstract postulate is necessary which is in fact capable of being proved from the other postulates that we have already on hand. The only effects of this excess of abstract postulates are to diminish our aesthetic pleasure in mathematical reasoning, and to give us more trouble when we come to the third process of criticism.

This third process of criticism is that of verifying that our abstract postulates hold for the particular case in question. It is in respect to this process of verification for the particular case that all the trouble arises. In some simple instances, such as the counting of forty apples, we can with a little care arrive at practical certainty. But in general, with more complex instances, complete certainty is unattainable. Volumes, libraries of volumes, have been written on the subject. It is the battle-ground of rival philosophers. There are two distinct questions involved. There are particular definite things observed, and we have to make sure that the relations between these things really do obey certain definite exact abstract conditions. There is great room for error here. The exact observational methods of science are all contrivances for limiting these erroneous conclusions as to direct matters of fact. But another question arises. The things directly observed are, almost always, only samples. We want to conclude that the abstract conditions, which hold for the samples, also hold for all other entities which, for some reason or other, appear to us to be of the same sort. This process of reasoning from the sample to the whole species is induction. The theory of induction is the despair of philosophy – and yet all our activities are based upon it. Anyhow, in criticizing a mathematical conclusion as to a particular matter of fact, the real difficulties consist in finding out the abstract assumptions involved, and in estimating the evidence for their applicability to the particular case in hand.

It often happens, therefore, that in criticizing a learned book of applied mathematics, or a memoir, one's whole trouble is with the first chapter, or even with the first page. For it is there, at the very outset, where the author will probably be found to slip in his assumptions. Further, the trouble is not with what the author does say, but with what he does not say. Also it is not with what he knows he has assumed, but with what he has unconsciously assumed. We do not doubt the author's honesty. It is his perspicacity which we are criticizing. Each generation criticizes the unconscious assumptions made by its parents. It may assent to them, but it brings them out into the open.

The history of the development of language illustrates this point. It is a history of the progressive analysis of ideas. Latin and Greek were inflected languages. This means that they express an unanalysed complex of ideas by the mere modification of a word; whereas in English, for example, we use prepositions and auxiliary verbs to drag into the open the whole bundle of ideas involved. For certain forms of literary art – though not always – the compact absorption of auxiliary ideas into the main word may be an advantage. But in a language such as English there is the overwhelming gain in explicitness. This increased explicitness is a more complete exhibition of the various abstractions involved in the complex idea which is the meaning of the sentence.

By comparison with language, we can now see what is the function in thought which is performed by pure mathematics. It is a resolute attempt to go the whole way in the direction of complete analysis, so as to separate the elements of mere matter of fact from the purely abstract conditions which they exemplify.

The habit of such analysis enlightens every act of the functioning of the human mind. It first (by isolating it) emphasizes the direct aesthetic appreciation of the content of experience. This direct appreciation means an apprehension of what this experience is in itself in its own particular essence, including its immediate concrete values. This is a question of direct experience, dependent upon sensitive subtlety. There is then the abstraction of the particular entities involved, viewed in themselves, and as apart from that particular occasion of experience in which we are then apprehending them. Lastly there is the further apprehension of the absolutely general conditions satisfied by the particular relations of those entities as in that experience. These conditions gain their generality from the fact that they are expressible without reference to those particular relations or to those particular *relata* which occur in that particular occasion of experience. They are conditions which might hold for an indefinite variety of other occasions, involving other entities and other relations between them. Thus these conditions are perfectly general because they refer to no particular occasion, and to no particular entities (such as green, or blue, or trees) which enter into a variety of occasions, and to no particular relationships between such entities.

There is, however, a limitation to be made to the generality of mathematics; it is a qualification which applies equally to all general statements. No statement, except one, can be made respecting any remote occasion which enters into no relationship with the immediate occasion so as to form a constitutive element of the essence of that immediate occasion. By the 'immediate occasion' I mean that occasion which involves as an ingredient the individual act of judgement in question. The one excepted statement is – If anything out of relationship, then complete ignorance as to it. Here by 'ignorance', I mean *ignorance*; accordingly no advice can be given as to how to expect it, or to treat it, in 'practice' or in any other way. Either we know something of the remote occasion by the cognition which is itself an element of the immediate occasion, or we know nothing. Accordingly the full universe, disclosed for every variety of experience, is a universe in which every detail enters into its proper relationship with the immediate occasion. The generality of mathematics is the most complete generality consistent with the community of occasions which constitutes our metaphysical situation.

It is further to be noticed that the particular entities require these general conditions for their ingression into any occasions; but the same general conditions may be required by many types of particular entities. This fact, that the general conditions transcend any one set of particular entities, is the ground for the entry into mathematics, and into mathematical logic, of the notion of the 'variable'. It is by the employment of this notion that general conditions are investigated without any specification of particular entities. This irrelevance of the particular entities has not been generally understood : for example, the shape-iness of shapes, e.g. circularity and sphericity and cubicality as in actual experience, do not enter into the geometrical reasoning.

The exercise of logical reason is always concerned with these absolutely general conditions. In its broadest sense, the discovery of mathematics is the discovery that the totality of these general abstract conditions, which are concurrently applicable to the relationships among the entities of any one concrete occasion, are themselves interconnected in the manner of a pattern with a key to it. This pattern of relationships among general abstract conditions is imposed alike on external reality, and on our abstract representations of it, by the general necessity that every thing must be just its own individual self, with its own individual way of differing from everything else. This is nothing else than the necessity of abstract logic, which is the presupposition involved in the very fact of interrelated existence as disclosed in each immediate occasion of experience.

The key to the patterns means this fact: that from a select set of those general conditions, exemplified in any one and the same occasion, a pattern involving an infinite variety of other such conditions, also exemplified in the same occasion, can be developed by the pure exercise of abstract logic. Any such select set is called the set of postulates, or premises, from which the reasoning proceeds. The reasoning is nothing else than the exhibition of the whole pattern of general conditions involved in the pattern derived from the selected postulates.

The harmony of the logical reason, which divines the complete pattern as involved in the postulates, is the most general aesthetic property arising from the mere fact of concurrent existence in the unity of one occasion. Wherever there is a unity of occasion there is thereby established an aesthetic relationship between the general conditions involved in that occasion. This aesthetic relationship is that which is divined in the exercise of rationality. Whatever falls within that relationship is thereby exemplified in that occasion, whatever falls without that relationship is thereby excluded from exemplification in that occasion. The complete pattern of general conditions, thus exemplified, is determined by any one of many select sets of these conditions. These key sets are sets of equivalent postulates. This reasonable harmony of being, which is required for the unity of a complex occasion, together with the completeness of the realization (in that occasion) of all that is involved in its logical harmony, is the primary article of metaphysical doctrine. It means that for things to be together involves that they are reasonably together. This means that thought can penetrate into every occasion of fact, so that by comprehending its key conditions, the whole complex of its pattern of conditions lies open before it. It comes to this: provided we know something which is perfectly general about the elements in any occasion, we can then know an indefinite number of other equally general concepts which must also be exemplified in that same occasion. The logical harmony involved in the unity of an occasion is both exclusive and inclusive. The occasion must exclude the inharmonious, and it must include the harmonious.

Pythagoras was the first man who had any grasp of the full sweep of this general principle. He lived in the sixth century before Christ. Our knowledge of him is fragmentary. But we know some points which establish his greatness in the history of thought. He insisted on the importance of the utmost generality in reasoning, and he divined the importance of number as an aid to the construction of any representation of the conditions involved in the order of nature. We know also that he studied geometry, and discovered the general proof of the remarkable theorem about right-angled triangles. The formation of the Pythagorean brotherhood, and the mysterious rumours as to its rites and its influence, afford some evidence that Pythagoras divined, however dimly, the possible importance of mathematics in the formation of science. On the side of philosophy he started a discussion which has agitated thinkers ever since. He asked, 'What is the status of mathematical entities, such as numbers for example, in the realm of things?' The number 'two', for example, is in some sense exempt from the flux of time and the necessity of position in space. Yet it is involved in the real world. The same considerations apply to geometrical notions – to circular shapes, for example.

Pythagoras is said to have taught that the mathematical entities, such as numbers and shapes, were the ultimate stuff out of which the real entities of our perceptual experience are constructed. As thus baldly stated, the idea seems crude, and indeed silly. But undoubtedly, he had hit upon a philosophical notion of considerable importance; a notion which has a long history, and which has moved the minds of men, and has even entered into Christian theology. About a thousand years separate the Athanasian Creed from Pythagoras, and about two thousand four hundred years separate Pythagoras from Hegel. Yet for all these distances in time, the importance of definite number in the constitution of the divine nature, and the concept of the real world as exhibiting the evolution of an idea, can both be traced back to the train of thought set going by Pythagoras.

The importance of an individual thinker owes something to chance. For it depends upon the fate of his ideas in the minds of his successors. In this respect Pythagoras was fortunate. His philosophical speculations reach us through the mind of Plato. The Platonic world of ideas is the refined, revised form of the Pythagorean doctrine that number lies at the base of the real world. Owing to the Greek mode of representing numbers by patterns of dots, the notions of number and of geometrical configuration are less separated than with us. Also Pythagoras, without doubt, included the shape-iness of shape, which is an impure mathematical entity. So today, when Einstein and his followers proclaim that physical facts, such as gravitation, are to be construed as exhibitions of local peculiarities of spatio-temporal properties, they are following the pure Pythagorean tradition. In a sense, Plato and Pythagoras stand nearer to modern physical science than does Aristotle. The two former were mathematicians, whereas Aristotle was the son of a doctor, though of course he was not thereby ignorant of mathematics. The practical counsel to be derived from Pythagoras is to measure, and thus to express quality in terms of numerically determined quantity. But the biological sciences, then and till our own time, have been overwhelmingly classificatory. Accordingly, Aristotle by his logic throws the emphasis on classification. The popularity of Aristotelian logic retarded the advance of physical science throughout the Middle Ages. If only the schoolmen had measured instead of classifying, how much they might have learnt!

Classification is a half-way house between the immediate concreteness of the individual thing and the complete abstraction of mathematical notions. The species take account of the specific character, and the genera of the generic character. But in the procedure of relating mathematical notions to the facts of nature, by counting, by measurement, and by geometrical relations, and by types of order, the rational contemplation is lifted from the incomplete abstractions involved in definite species and genera, to the complete abstractions of mathematics. Classification is necessary. But unless you can progress from classification to mathematics, your reasoning will not take you very far.

Between the epoch which stretches from Pythagoras to Plato and the epoch comprised in the seventeenth century of the modern world nearly two thousand years elapsed. In this long interval mathematics had made immense strides. Geometry had gained the study of conic sections and trigonometry; the method of exhaustion had almost anticipated the integral calculus; and above all the Arabic arithmetical notation and algebra had been contributed by Asiatic thought. But the progress was on technical lines. Mathematics, as a formative element in the development of philosophy, never, during this long period, recovered from its deposition at the hands of Aristotle. Some of the old ideas derived from the Pythagorean-Platonic epoch lingered on, and can be traced among the Platonic influences which shaped the first period of evolution of Christian theology. But philosophy received no fresh inspiration from the steady advance of mathematical science. In the seventeenth century the influence of Aristotle was at its lowest, and mathematics recovered the importance of its earlier period. It was an age of great physicists and great philosophers; and the physicists and philosophers were alike mathematicians. The exception of John Locke should be made; although he was greatly influenced by the Newtonian circle of the Royal Society. In the age of Galileo, Descartes, Spinoza, Newton, and Leibniz, mathematics was an influence of the first magnitude in the formation of philosophic ideas. But the mathematics, which now emerged into prominence, was a very different science from the mathematics of the earlier epoch. It had gained in generality, and had started upon its almost incredible modern career of piling subtlety of generalization upon subtlety of generalization; and of finding, with each growth of complexity, some new application, either to physical science or to philosophic thought. The Arabic notation had equipped the science with almost perfect technical efficiency in the manipulation of numbers. This relief from a struggle with arithmetical details (as instanced, for example, in the Egyptian arithmetic of 1600 BC) gave room for a development which had already been faintly anticipated in later Greek mathematics. Algebra now came upon the scene, and algebra is a generalization of arithmetic. In the same way as the notion of number abstracted from reference to any one particular set of entities, so in algebra abstraction is made from the notion of any particular numbers. Just as the number '5' refers impartially to any group of five entities, so in algebra the letters are used to refer impartially to any number, with the proviso that each letter is to refer to the same number throughout the same context of its employment.

This usage was first employed in equations, which are methods of asking complicated arithmetical questions. In this connection, the letters representing numbers were termed 'unknowns'. But equations soon suggested a new idea, that, namely, of a function of one or more general symbols, these symbols being letters representing any numbers. In this employment the algebraic letters are called the 'arguments' of the function, or sometimes they are called the 'variables'. Then, for instance, if an angle is represented by an algebraical letter, as standing for its numerical measure in terms of a given unit, trigonometry is absorbed into this new algebra. Algebra thus develops into the general science of analysis in which we consider the properties of various functions of undetermined arguments. Finally the particular functions, such as the trigonometrical functions, and the logarithmic functions, and the algebraic functions, are generalized into the idea of 'any function'. Too large a generalization leads to mere barrenness. It is the large generalization, limited by a happy particularity, which is the fruitful conception. For instance the idea of any *continuous* function, whereby the limitation of continuity is introduced, is the fruitful idea which has led to most of the important applications. This rise of algebraic analysis was concurrent with Descartes' discovery of analytical geometry, and then with the invention of the infinitesimal calculus by Newton and Leibniz. Truly, Pythagoras, if he could have foreseen the issue of the train of thought which he had set going would have felt himself fully justified in his brotherhood with its excitement of mysterious rites.

The point which I now want to make is that this dominance of the idea of functionality in the abstract sphere of mathematics found itself reflected in the order of nature under the guise of mathematically expressed laws of nature. Apart from this progress of mathematics, the seventeenth-century developments of science would have been impossible. Mathematics supplied the background of imaginative thought with which the men of science approached the observation of nature. Galileo produced formulae, Descartes produced formulae, Huyghens produced formulae, Newton produced formulae.

As a particular example of the effect of the abstract development of mathematics upon the science of those times, consider the notion of periodicity. The general recurrences of things are very obvious in our ordinary experience. Days recur, lunar phases recur, the seasons of the year recur, rotating bodies recur to their old positions, beats of the heart recur, breathing recurs. On every side, we are met by recurrence. Apart from recurrence, knowledge would be impossible; for nothing could be referred to our past experience. Also, apart from some regularity of recurrence, measurement would be impossible. In our experience, as we gain the idea of exactness, recurrence is fundamental.

In the sixteenth and seventeenth centuries, the theory of periodicity took a fundamental place in science. Kepler divined a law connecting the major axes of the planetary orbits with the periods in which the planets respectively described their orbits : Galileo observed the periodic vibrations of pendulums : Newton explained sound as being due to the disturbance of air by the passage through it of periodic waves of condensation and rarefaction : Huyghens explained light as being due to the transverse waves of vibration of a subtle ether : Mersenne connected the period of the vibration of a violin string with its density, tension, and length. The birth of modern physics depended upon the application of the abstract idea of periodicity to a variety of concrete instances. But this would have been impossible, unless mathematicians had already worked out in the abstract the various abstract ideas which cluster round the notions of periodicity. The science of trigonometry arose from that of the relations of the angles of a right-angled triangle, to the ratios between the sides and hypotenuse of the triangle. Then, under the influence of the newly discovered mathematical science of the analysis of functions, it broadened out into the study of the simple abstract periodic functions which these ratios exemplify. Thus trigonometry became completely abstract; and in thus becoming abstract, it became useful. It illuminated the underlying analogy between sets of utterly diverse physical phenomena; and at the same time it supplied the weapons by which any one such set could have its various features analysed and related to each other.[1]

Nothing is more impressive than the fact that as mathematics withdrew increasingly into the upper regions of ever greater extremes of abstract thought, it returned back to earth with a corresponding growth of importance for the analysis of concrete fact. The history of the seventeenth-century science reads as though it were some vivid dream of Plato or Pythagoras. In this characteristic the seventeenth century was only the forerunner of its successors.

The paradox is now fully established that the utmost abstractions are the true weapons with which to control our thought of concrete fact. As the result of the prominence of mathematics in the seventeenth century, the eighteenth century was mathematically minded, more especially where French influence predominated. An exception must be made of the English empiricism derived from Locke. Outside France, Newton's direct influence on philosophy is best seen in Kant, and not in Hume.

In the nineteenth century, the general influence of mathematics waned. The romantic movement in literature, and the idealistic movement in philosophy were not the products of mathematical minds. Also, even in science, the growth of geology, of zoology, and of the biological sciences generally, was in each case entirely disconnected from any reference to mathematics. The chief scientific excitement of the century was the Darwinian theory of evolution. Accordingly, mathematicians were in the background so far as the general thought of that age was concerned. But this does not mean that mathematics was being neglected, or even that it was uninfluential. During the nineteenth century pure mathematics made almost as much progress as during all the preceding centuries from Pythagoras onwards. Of course progress was easier, because the technique had been per-

[1] For a more detailed consideration of the nature and function of pure mathematics cf. my *Introduction to Mathematics*, Home University Library, Williams and Norgate, London.

fected. But allowing for that, the change in mathematics between the years 1800 and 1900 is very remarkable. If we add in the previous hundred years, and take the two centuries preceding the present time, one is almost tempted to date the foundation of mathematics somewhere in the last quarter of the seventeenth century. The period of the discovery of the elements stretches from Pythagoras to Descartes, Newton, and Leibniz, and the developed science has been created during the last two hundred and fifty years. This is not a boast as to the superior genius of the modern world; for it is harder to discover the elements than to develop the science.

Throughout the nineteenth century, the influence of the science was its influence on dynamics and physics, and thence derivatively on engineering and chemistry. It is difficult to overrate its indirect influence on human life through the medium of these sciences. But there was no direct influence of mathematics upon the general thought of the age.

In reviewing this rapid sketch of the influence of mathematics throughout European history, we see that it had two great periods of direct influence upon general thought, both periods lasting for about two hundred years. The first period was that stretching from Pythagoras to Plato, when the possibility of the science, and its general character, first dawned upon the Grecian thinkers. The second period comprised the seventeenth and eighteenth centuries of our modern epoch. Both periods had certain common characteristics. In the earlier, as in the later period, the general categories of thought in many spheres of human interest, were in a state of disintegration. In the age of Pythagoras, the unconscious Paganism, with its traditional clothing of beautiful ritual and of magical rites, was passing into a new phase under two influences. There were waves of religious enthusiasm, seeking direct enlightenment into the secret depths of being; and at the opposite pole, there was the awakening of critical analytical thought, probing with cool dispassionateness into ultimate meanings. In both influences, so diverse in their outcome, there was one common element – an awakened curiosity, and a movement towards the reconstruction of traditional ways. The pagan mysteries may be compared to the Puritan reaction and to the Catholic reaction; critical scientific interest was alike in both epochs, though with minor differences of substantial importance.

In each age, the earlier stages were placed in periods of rising prosperity, and of new opportunities. In this respect, they differed from the period of gradual declension in the second and third centuries when Christianity was advancing to the conquest of the Roman world. It is only in a period, fortunate both in its opportunities for disengagement from the immediate pressure of circumstances, and in its eager curiosity, that the age-spirit can undertake any direct revision of those final abstractions which lie hidden in the more concrete concepts from which the serious thought of an age takes its start. In the rare periods when this task can be undertaken, mathematics becomes relevant to philosophy. For mathematics is the science of the most complete abstractions to which the human mind can attain.

The parallel between the two epochs must not be pressed too far. The modern world is larger and more complex than the ancient civilization round the shores of the Mediterranean, or even than that of the Europe which sent Columbus and the Pilgrim Fathers across the ocean. We cannot now explain our age by some simple formula which becomes dominant and will then be laid to rest for a thousand years. Thus the temporary submergence of the mathematical mentality from the time of Rousseau onwards appears already to be at an end. We are entering upon an age of reconstruction, in religion, in science, and in political thought. Such ages, if they are to avoid mere ignorant oscillation between extremes, must seek truth in its ultimate depths. There can be no vision of this depth of truth apart from a philosophy which takes full account of those ultimate abstractions, whose interconnections it is the business of mathematics to explore.

In order to explain exactly how mathematics is gaining in general importance at the present time, let us start from a particular scientific perplexity and consider the notions to which we are naturally led by some attempt to unravel its difficulties. At present physics is troubled by the quantum theory. I need not now explain[2] what this theory is, to those who are not already familiar with it. But the point is that one of the most hopeful lines of explanation is to assume that an electron does not continuously traverse its path in space. The alternative notion as to its mode of existence is that it appears at a series of discrete positions in space which it occupies for successive durations of time. It is as though an automobile moving at the average rate of thirty miles an hour along a road, did not traverse the road continuously; but appeared successively at the successive milestones, remaining for two minutes at each milestone.

In the first place there is required the purely technical use of mathematics to determine whether this conception does in fact explain the many perplexing characteristics of the quantum theory. If the notion survives this test, undoubtedly physics will adopt it. So far the question is purely one for mathematics and physical science to settle between them, on the basis of mathematical calculations and physical observations.

But now a problem is handed over to the philosophers. This discontinuous existence in space, thus assigned to electrons, is very unlike the continuous existence of material entities which we habitually assume as obvious. The electron seems to be borrowing the character which some people have assigned to the Mahatmas of Tibet. These electrons, with the correlative protons, are now conceived as being the fundamental entities out of which the material bodies of ordinary experience are composed. Accordingly

[2] Cf. Chapter VIII.

if this explanation is allowed, we have to revise all our notions of the ultimate character of material existence. For when we penetrate to these final entities, this startling discontinuity of spatial existence discloses itself.

There is no difficulty in explaining the paradox, if we consent to apply to the apparently steady undifferentiated endurance of matter the same principles as those now accepted for sound and light. A steadily sounding note is explained as the outcome of vibrations in the air : a steady colour is explained as the outcome of vibrations in ether. If we explain the steady endurance of matter on the same principle, we shall conceive each primordial element as a vibratory ebb and flow of an underlying energy, or activity. Suppose we keep to the physical idea of energy : then each primordial element will be an organized system of vibratory streaming of energy. Accordingly there will be a definite period associated with each element; and within that period the stream-system will sway from one stationary maximum to another stationary maximum – or, taking a metaphor from the ocean tides, the system will sway from one high tide to another high tide. This system, forming the primordial element, is nothing at any instant. It requires its whole period in which to manifest itself. In an analogous way, a note of music is nothing at an instant, but it also requires its whole period in which to manifest itself.

Accordingly, in asking where the primordial element is, we must settle on its average position at the centre of each period. If we divide time into smaller elements, the vibratory system as one electronic entity has no existence. The path in space of such a vibratory entity – where the entity is *constituted by* the vibrations – must be represented by a series of detached positions in space, analogously to the automobile which is found at successive milestones and at nowhere between.

We first must ask whether there is any evidence to associate the quantum theory with vibration. This question is immediately answered in the affirmative. The whole theory centres round the radiant energy from an atom, and is intimately associated with the periods of the radiant wave-systems. It seems, therefore, that the hypothesis of essentially vibratory existence is the most hopeful way of explaining the paradox of the discontinuous orbit.

In the second place, a new problem is now placed before philosophers and physicists, if we entertain the hypothesis that the ultimate elements of matter are in their essence vibratory. By this I mean that apart from being a periodic system, such an element would have no existence. With this hypothesis we have to ask, what are the ingredients which form the vibratory organism? We have already got rid of the matter with its appearance of undifferentiated endurance. Apart from some metaphysical compulsion, there is no reason to provide another more subtle stuff to take the place of the matter which has just been explained away. The field is now open for the introduction of some new doctrine of organism which may take the place of the materialism with which, since the seventeenth century, science has saddled philosophy. It must be remembered that the physicists' energy is obviously an abstraction. The concrete fact, which is the organism, must be a complete expression of the character of a real occurrence. Such a displacement of scientific materialism, if it ever takes place, cannot fail to have important consequences in every field of thought.

Finally, our last reflection must be, that we have in the end come back to a version of the doctrine of old Pythagoras, from whom mathematics, and mathematical physics, took their rise. He discovered the importance of dealing with abstractions; and in particular directed attention to number as characterizing the periodicities of notes of music. The importance of the abstract idea of periodicity was thus present at the very beginning both of mathematics and of European philosophy.

In the seventeenth century, the birth of modern science required a new mathematics, more fully equipped for the purpose of analysing the characteristics of vibratory existence. And now in the twentieth century we find physicists largely engaged in analysing the periodicities of atoms. Truly, Pythagoras in founding European philosophy and European mathematics, endowed them with the luckiest of lucky guesses – or, was it a flash of divine genius, penetrating to the inmost nature of things?

CHAPTER III

THE CENTURY OF GENIUS

The previous chapters were devoted to the antecedent conditions which prepared the soil for the scientific outburst of the seventeenth century. They traced the various elements of thought and instinctive belief, from their first efflorescence in the classical civilization of the ancient world, through the transformations which they underwent in the Middle Ages, up to the historical revolt of the sixteenth century. Three main factors arrested attention – the rise of mathematics, the instinctive belief in a detailed order of nature and the unbridled rationalism of the thought of the later Middle Ages. By this rationalism I mean the belief that the avenue to truth was predominantly through a metaphysical analysis of the nature of things, which would thereby determine how things acted and functioned. The historical revolt was the definite abandonment of this method in favour of the study of the empirical fact of antecedents and consequences. In religion, it meant the appeal to the origins of Christianity; and in science it meant the appeal to experiment and the inductive method of reasoning.

A brief, and sufficiently accurate, description of the intellectual life of the European races during the succeeding two centuries and a quarter up to our own times is that they have been living upon the accumulated capital of ideas provided for them by the genius of the seventeenth century. The men of this epoch inherited a ferment of ideas attendant upon the historical revolt of the sixteenth century, and they bequeathed formed systems of thought touching every aspect of human life. It is the one century which consistently, and throughout the whole range of human activities, provided intellectual genius adequate for the greatness of its occasions. The crowded stage of this hundred years is indicated by the coincidences which mark its literary annals. At its dawn Bacon's *Advancement of Learning* and Cervantes' *Don Quixote* were published in the same year (1605), as though the epoch would introduce itself with a forward and a backward glance. The first quarto edition of *Hamlet* appeared in the preceding year, and a slightly variant edition in the same year. Finally Shakespeare and Cervantes died on the same day, 23 April 1616. In the spring of this same year Harvey is believed to have first expounded his theory of the circulation of the blood in a course of lectures before the College of Physicians in London. Newton was born in the year that Galileo died (1642), exactly one hundred years after the publication of Copernicus' *De Revolutionibus*. One year earlier Descartes published his *Meditationes* and two years later his *Principia Philosophiae*. There simply was not time for the century to space out nicely its notable events concerning men of genius.

I cannot now enter upon a chronicle of the various stages of intellectual advance included within this epoch. It is too large a topic for one lecture, and would obscure the ideas which it is my purpose to develop. A mere rough catalogue of some names will be sufficient, names of men who published to the world important work within these limits of time: Francis Bacon, Harvey, Kepler, Galileo, Descartes, Pascal, Huyghens, Boyle, Newton, Locke, Spinoza, Leibniz. I have limited the list to the sacred number of twelve, a number much too small to be properly representative. For example, there is only one Italian there, whereas Italy could have filled the list from its own ranks. Again Harvey is the only biologist, and also there are too many Englishmen. This latter defect is partly due to the fact that the lecturer is English, and that he is lecturing to an audience which, equally with him, owns this English century. If he had been Dutch, there would have been too many Dutchmen; if Italian, too many Italians; and if French, too many Frenchmen. The unhappy Thirty Years' War was devastating Germany; but every other country looks back to this century as an epoch which witnessed some culmination of its genius. Certainly this was a great period of English thought; as at a later time Voltaire impressed upon France.

The omission of physiologists, other than Harvey, also requires explanation. There were, of course, great advances in biology within the century, chiefly associated with Italy and the University of Padua. But my purpose is to trace the philosophic outlook, derived from science and presupposed by science, and to estimate some of its effects on the general climate of each age. Now the scientific philosophy of this age was dominated by physics; so as to be the most obvious rendering, in terms of general ideas, of the state of physical knowledge of that age and of the two succeeding centuries. As a matter of fact, these concepts are very unsuited to biology; and set for it an insoluble problem of matter and life and organism, with which biologists are now wrestling. But the science of living organisms is only now coming to a growth adequate to impress its conceptions upon philosophy. The last half century before the present time has witnessed unsuccessful attempts to impress biological notions upon the materialism of the seventeenth century. However this success be estimated, it is certain that the root ideas of the seventeenth century were derived from the school of thought which produced Galileo, Huyghens and Newton, and not from the physiologists of Padua. One unsolved problem of thought, so far as it derives from this period, is to be formulated thus: given configurations of matter with locomotion in space as assigned by physical laws, to account for living organisms.

My discussion of the epoch will be best introduced by a quotation from Francis Bacon, which forms the opening of Section (or 'Century') IX of his *Natural History*, I mean his *Silva Silvarum*. We are told in the contemporary memoir by his chaplain, Dr Rawley, that this work was composed in the last five years of his life, so it must be dated between 1620 and 1626. The quotation runs thus:

> It is certain that all bodies whatsoever, though they have no sense, yet they have perception: for when one body is applied to another, there is a kind of election to embrace that which is agreeable, and to exclude or expel that which is ingrate; and whether the body be alterant or altered, evermore a perfection precedeth operation; for else all bodies would be alike one to another. And sometimes this perception, in some kind of bodies, is far more subtile than sense; so that sense is but a dull thing in comparison of it: we see a weatherglass will find the least difference of the weather in heat or cold, when we find it not. And this perception is sometimes at a distance, as well as upon the touch; as when the loadstone draweth iron; or flame naphtha of Babylon, a great distance off. It is therefore a subject of a very noble inquiry, to inquire of the more subtile perceptions; for it is another key to open nature, as well as the sense; and sometimes better. And besides, it is a principal means of natural divination; for that which in these perceptions appeareth early, in the great effects cometh long after.

There are a great many points of interest about this quotation, some of which will emerge into importance in succeeding lectures. In the first place, note the careful way in which Bacon discriminates between *perception*, or *taking account of*, on the one hand, and *sense*, or *cognitive experience*, on the other hand. In this respect Bacon is outside the physical line of thought which finally dominated the century. Later on, people thought of passive matter which was operated on externally by forces. I believe Bacon's line of thought to have expressed a more fundamental truth than do the materialistic concepts which were then being shaped as adequate for physics. We are now so used to the materialistic way of looking at things, which has been rooted in our literature by the genius of the seventeenth century, that it is with some difficulty that we understand the possibility of another mode of approach to the problems of nature.

In the particular instance of the quotation which I have just made, the whole passage and the context in which it is embedded, are permeated through and through by the experimental method, that is to say, by attention to 'irreducible and stubborn facts', and by the inductive method of eliciting general laws. Another unsolved problem which has been bequeathed to us by the seventeenth century is the rational justification of this method of induction. The explicit realization of the antithesis between the deductive rationalism of the scholastics and the inductive observational methods of the moderns must chiefly be ascribed to Bacon; though, of course, it was implicit in the mind of Galileo and of all the men of science of those times. But Bacon was one of the earliest of the whole group, and also had the most direct apprehension of the full extent of the intellectual revolution which was in progress. Perhaps the man who most completely anticipated both Bacon and the whole modern point of view was the artist Leonardo da Vinci, who lived almost exactly a century before Bacon. Leonardo also illustrated the theory which I was advancing in my last lecture, that the rise of naturalistic art was an important ingredient in the formation of our scientific mentality. Indeed, Leonardo was more completely a man of science than was Bacon. The practice of naturalistic art is more akin to the practice of physics, chemistry and biology than is the practice of law. We all remember the saying of Bacon's contemporary, Harvey, the discoverer of the circulation of the blood, that Bacon 'wrote of science like a Lord Chancellor'. But at the beginning of the modern period da Vinci and Bacon stand together as illustrating the various strains which have combined to form the modern world, namely, legal mentality and the patient observational habits of the naturalistic artists.

In the passage which I have quoted from Bacon's writings there is no explicit mention of the method of inductive reasoning. It is unnecessary for me to prove to you by any quotations that the enforcement of the importance of this method, and of the importance to the welfare of mankind, the secrets of nature to be thus discovered, was one of the main themes to which Bacon devoted himself in his writings. Induction has proved to be a somewhat more complex process than Bacon anticipated. He had in his mind the belief that with a sufficient care in the collection of instances the general law would stand out of itself. We know now, and probably Harvey knew then, that this is a very inadequate account of the processes which issue in scientific generalizations. But when you have made all the requisite deductions, Bacon remains as one of the great builders who constructed the mind of the modern world.

The special difficulties raised by induction emerged in the eighteenth century, as the result of Hume's criticism. But Bacon was one of the prophets of the historical revolt, which deserted the method of unrelieved rationalism, and rushed into the other extreme of basing all fruitful knowledge upon inference from particular occasions in the past to particular occasions in the future. I do not wish to throw any doubt upon the validity of induction, when it has been properly guarded. My point is, that the very baffling task of applying reason to elicit the general characteristics of the immediate occasion, as set before us in direct cognition, is a necessary preliminary, if we are to justify induction; unless indeed we are content to base it upon our vague instinct that of course it is all right. Either there is something about the immediate occasion which affords knowledge of the past and the future, or we are reduced to utter scepticism as to memory and induction. It is impossible to overemphasize the point that the key to the process of induction, as used either in science or in our ordinary life, is to be found in the right understanding of the immediate occasion of knowledge in its full concreteness. It is in respect to our grasp of the character of these occasions in their concreteness that the modern developments of physiology and of psychology are of critical importance. I shall illustrate this point in my subsequent lectures. We find ourselves amid insoluble difficulties when we substitute for this concrete occasion a mere abstract in which we only consider material objects in a flux of configurations in time and space. It is quite obvious that such objects can tell us only that they are where they are.

Accordingly, we must recur to the method of the school-divinity as explained by the Italian medievalists whom I quoted in the first lecture. We must observe the immediate occasion, and use reason to elicit a general description of its nature. Induction presupposes metaphysics. In other words, it rests upon an antecedent rationalism. You cannot have a rational justification for your appeal to history till your metaphysics has assured you that there *is* a history to appeal to; and likewise your conjectures as to the future presuppose some basis of knowledge that there *is* a future already subjected to some determinations. The difficulty is to make sense of either of these ideas. But unless you have done so, you have made nonsense of induction.

You will observe that I do not hold induction to be in its essence the derivation of general laws. It is the divination of some characteristics of a particular future from the known characteristics of a particular past. The wider assumption of general laws holding for all cognizable occasions appears a very unsafe addendum to attach to this limited knowledge. All we can ask of the present occasion is that it shall determine a particular community of occasions, which are in some respects mutually qualified by reason of their inclusion within that same community. That community of occasions considered in physical science is the set of happenings which fit on to each other – as we say – in a common space-time, so that we can trace the transitions from one to the other. Accordingly, we refer to *the* common space-time indicated in our immediate occasion of knowledge. Inductive reasoning proceeds from the particular occasion to the particular community of occasions, and from the particular community to relations between particular occasions within that community. Until we have taken into account other scientific concepts, it is impossible to carry the discussion of induction further than this preliminary conclusion.

The third point to notice about this quotation from Bacon is the purely qualitative character of the statements made in it. In this respect Bacon completely missed the tonality which lay behind the success of seventeenth-century science. Science was becoming, and has remained, primarily quantitative. Search for measurable elements among your phenomena, and then search for relations between these measures of physical quantities. Bacon ignores this rule of science. For example, in the quotation given he speaks of action at a distance; but he is thinking qualitatively and not quantitatively. We cannot ask that he should anticipate his younger contemporary Galileo, or his distant successor Newton. But he gives no hint that there should be a search for quantities. Perhaps he was misled by the current logical doctrines which had come down from Aristotle. For, in effect, these doctrines said to the physicist 'classify' when they should have said 'measure'.

By the end of the century physics had been founded on a satisfactory basis of measurement. The final and adequate exposition was given by Newton. The common measurable element of mass was discerned as characterizing all bodies in different amounts. Bodies which are apparently identical in substance, shape, and size have very approximately the same mass : the closer the identity, the nearer the equality. The force acting on a body, whether by touch or by action at a distance, was (in effect) defined as being equal to the mass of the body multiplied by the rate of change of the body's velocity, so far as this rate of change is produced by that force. In this way the force is discerned by its effect on the motion of the body. The question now arises whether this conception of the magnitude of a force leads to the discovery of simple quantitative laws involving the alternative determination of forces by circumstances of the configuration of substances and of their physical characters. The Newtonian conception has been brilliantly successful in surviving this test throughout the whole modern period. Its first triumph was the law of gravitation. Its cumulative triumph has been the whole development of dynamical astronomy, of engineering, and of physics.

This subject of the formation of the three laws of motion and of the law of gravitation deserves critical attention. The whole development of thought occupied exactly two generations. It commenced with Galileo and ended with Newton's *Principia*; and Newton was born in the year that Galileo died. Also the lives of Descartes and Huyghens fall within the period occupied by these great terminal figures. The issue of the combined labours of these four men has some right to be considered as the greatest single intellectual success which mankind has achieved. In estimating its size, we must consider the completeness of its range. It constructs for us a vision of the material universe, and it enables us to calculate the minutest detail of a particular occurrence. Galileo took the first step in hitting on the right line of thought. He noted that the critical point to attend to was not the motion of bodies but the changes of their motions. Galileo's discovery is formularized by Newton in his first law of motion: 'Every body continues in its state of rest, or of uniform motion in a straight line, except so far as it may be compelled by force to change that state.'

This formula contains the repudiation of a belief which had blocked the progress of physics for two thousand years. It also deals with a fundamental concept which is essential to scientific theory; I mean, the concept of an ideally isolated system. This conception embodies a fundamental character of things, without which science, or indeed any knowledge on the part of finite intellects, would be impossible. The 'isolated' system is not a solipsist system, apart from which there would be nonentity. It is isolated as within the universe. This means that there are truths respecting this system which require reference only to the remainder of things by way of a uniform systematic scheme of relationships. Thus the conception of an isolated system is not the conception of substantial independence from the remainder of things, but of freedom from casual contingent dependence upon detailed items within the rest of the universe. Further, this freedom from casual dependence is required only in respect to certain abstract characteristics which attach to the isolated system, and not in respect to the system in its full concreteness.

The first law of motion asks what is to be said of a dynamically isolated system so far as concerns its motion as a whole, abstracting from its orientation and its internal arrangement of parts. Aristotle said that you must conceive such a system to be at rest. Galileo added that the state of rest is only a particular case, and that the general statement is 'either in a state of rest, or of uniform motion in a straight line'. Accordingly, an Aristotelian would conceive the forces arising from the reaction of alien bodies as being quantitatively measurable in terms of the velocity they sustain, and as directly determined by the direction of that velocity; while the Galilean would direct attention to the magnitude of the acceleration and to its direction. This difference is illustrated by contrasting Kepler and Newton. They both speculated as to the forces sustaining the planets in their orbits. Kepler looked for tangential forces pushing the planets along, whereas Newton looked for radial forces diverting the directions of the planets' motions.

Instead of dwelling upon the mistake which Aristotle made, it is more profitable to emphasize the justification which he had for it, if we consider the obvious facts of our experience. All the motions which enter into our normal everyday experience cease unless they are evidently sustained from the outside. Apparently, therefore, the sound empiricist must devote his attention to this question of the sustenance of motion. We here hit upon one of the dangers of unimaginative empiricism. The seventeenth century exhibits another example of this same danger; and, of all people in the world, Newton fell into it. Huyghens had produced the wave theory of light. But this theory failed to account for the most obvious facts about light as in our ordinary experience, namely, that shadows cast by obstructing objects are defined by rectilinear rays. Accordingly, Newton rejected this theory and adopted the corpuscular theory which completely explained shadows. Since then both theories have had their periods of triumph. At the present moment the scientific world is seeking for a combination of the two. These examples illustrate the danger of refusing to entertain an idea because of its failure to explain one of the most obvious facts in the subject matter in question. If you have had your attention directed to the novelties in thought in your own lifetime, you will have observed that almost all really new ideas have a certain aspect of foolishness when they are first produced.

Returning to the laws of motion, it is noticeable that no reason was produced in the seventeenth century for the Galilean as distinct from the Aristotelian position. It was an ultimate fact. When in the course of these lectures we come to the modern period, we shall see that the theory of relativity throws complete light on this question; but only by rearranging our whole ideas as to space and time.

It remained for Newton to direct attention to mass as a physical quantity inherent in the nature of a material body. Mass remained permanent during all changes of motion. But the proof of the permanence of mass amid chemical transformations had to wait for Lavoisier, a century later. Newton's next task was to find some estimate of the magnitude of the alien force in terms of the mass of the body and of its acceleration. He here had a stroke of luck. For, from the point of view of a mathematician, the simplest possible law, namely the product of the two, proved to be the successful one. Again the modern relativity theory modifies this extreme simplicity. But luckily for science the delicate experiments of the physicists of today were not then known, or even possible. Accordingly, the world was given the two centuries which it required in order to digest Newton's laws of motion.

Having regard to this triumph, can we wonder that scientists placed their ultimate principles upon a materialistic basis, and thereafter ceased to worry about philosophy? We shall grasp the course of thought, if we understand exactly what this basis is, and what difficulties it finally involves. When you are criticizing the philosophy of an epoch, do not chiefly direct your attention to those intellectual positions which its exponents feel it necessary explicitly to defend. There will be some fundamental assumptions which adherents of all the variant systems within the epoch unconsciously presuppose. Such assumptions appear so obvious that people do not know what they are assuming because no other way of putting things has ever occurred to them. With these assumptions a certain limited number of types of philosophic systems are possible, and this group of systems constitutes the philosophy of the epoch.

One such assumption underlies the whole philosophy of nature during the modern period. It is embodied in the conception which is supposed to express the most concrete aspect of nature. The Ionian philosophers asked, What is nature made of? The answer is couched in terms of stuff, or matter, or material – the particular name chosen is indifferent – which has the property of simple location in space and time, or, if you adopt the more modern ideas, in space-time. What I mean by matter, or material, is anything which has this property of simple location. By simple location I mean one major characteristic which refers equally both to space and time, and other minor characteristics which are diverse as between space and time.

The characteristic common both to space and time is

that material can be said to be *here* in space and *here* in time, or *here* in space-time, in a perfectly definite sense which does not require for its explanation any reference to other regions of space-time. Curiously enough this character of simple location holds whether we look on a region of space-time as determined absolutely or relatively. For if a region is merely a way of indicating a certain set of relations to other entities, then this characteristic, which I call simple location, is that material can be said to have just these relations of position to the other entities without requiring for its explanation any reference to other regions constituted by analogous relations of position to the same entities. In fact, as soon as you have settled, however you do settle, what you mean by a definite place in space-time, you can adequately state the relation of a particular material body to space-time by saying that it is just there, in that place; and, so far as simple location is concerned, there is nothing more to be said on the subject.

There are, however, some subordinate explanations to be made which bring in the minor characteristics which I have already mentioned. First, as regards time, if material has existed during any period, it has equally been in existence during any portion of that period. In other words, dividing the time does not divide the material. Secondly, in respect to space, dividing the volume does not divide the material. Accordingly, if material exists throughout a volume, there will be less of that material distributed through any definite half of that volume. It is from this property that there arises our notion of density at a point of space. Anyone who talks about density is not assimilating time and space to the extent that some extremists of the modern school of relativists very rashly desire. For the division of time functions, in respect to material, quite differently from the division of space.

Furthermore, this fact that the material is indifferent to the division of time leads to the conclusion that the lapse of time is an accident, rather than of the essence, of the material. The material is fully itself in any sub-period however short. Thus the transition of time has nothing to do with the character of the material. The material is equally itself at an instant of time. Half an instant of time is conceived as in itself without transition, since the temporal transition is the succession of instants.

The answer, therefore, which the seventeenth century gave to the ancient question of the Ionian thinkers, 'What is the world made of?' was that the world is a succession of instantaneous configurations of matter – or of material, if you wish to include stuff more subtle than ordinary matter, the ether for example.

We cannot wonder that science rested content with this assumption as to the fundamental elements of nature. The great forces of nature, such as gravitation, were entirely determined by the configurations of masses. Thus the configurations determined their own changes, so that the circle of scientific thought was completely closed. This is the famous mechanistic theory of nature, which has reigned supreme ever since the seventeenth century. It is the orthodox creed of physical science. Furthermore, the creed justified itself by the pragmatic test. It worked. Physicists took no more interest in philosophy. They emphasized the anti-rationalism of the historical revolt. But the difficulties of this theory of materialistic mechanism very soon became apparent. This history of thought in the eighteenth and nineteenth centuries is governed by the fact that the world had got hold of a general idea which it could neither live with nor live without.

This simple location of instantaneous material configurations is what Bergson has protested against, so far as it concerns time and so far as it is taken to be the fundamental fact of concrete nature. He calls it a distortion of nature due to the intellectual 'spatialization' of things. I agree with Bergson in his protest: but I do not agree that such distortion is a vice necessary to the intellectual apprehension of nature. I shall in subsequent lectures endeavour to show that this spatialization is the expression of more concrete facts under the guise of very abstract logical constructions. There is an error; but it is merely the accidental error of mistaking the abstract for the concrete. It is an example of what I will call the 'fallacy of misplaced concreteness'. This fallacy is the occasion of great confusion in philosophy. It is not necessary for the intellect to fall into the trap, though in this example there has been a very general tendency to do so.

It is at once evident that the concept of simple location is going to make great difficulties for induction. For, if in the location of configurations of matter throughout a stretch of time there is no inherent reference to any other times, past or future, it immediately follows that nature within any period does not refer to nature at any other period. Accordingly, induction is not based on anything which can be observed as inherent in nature. Thus we cannot look to nature for the justification of our belief in any law such as the law of gravitation. In other words, the order of nature cannot be justified by the mere observation of nature. For there is nothing in the present fact which inherently refers either to the past or to the future. It looks, therefore, as though memory, as well as induction, would fail to find any justification within nature itself.

I have been anticipating the course of future thought, and have been repeating Hume's argument. This train of thought follows so immediately from the consideration of simple location, that we cannot wait for the eighteenth century before considering it. The only wonder is that the world did in fact wait for Hume before noting the difficulty. Also it illustrates the anti-rationalism of the scientific public that, when Hume did appear, it was only the religious implications of his philosophy which attracted attention. This was because the clergy were in principle rationalists, whereas the men of science were content with a simple faith in the order of nature. Hume himself remarks, no doubt scoffingly, 'Our holy religion is founded on faith.'

This attitude satisfied the Royal Society but not the Church. It also satisfied Hume and has satisfied subsequent empiricists.

There is another presupposition of thought which must be put beside the theory of simple location. I mean the two correlative categories of substance and quality. There is, however, this difference. There were different theories as to the adequate description of the status of space. But whatever its status, no one had any doubt but that the connection with space enjoyed by entities, who are said to be in space, is that of simple location. We may put this shortly by saying that it was tacitly assumed that space is the locus of simple locations. Whatever is in space is *simpliciter* in some definite portion of space. But in respect to substance and quality the leading minds of the seventeenth century were definitely perplexed; though, with their usual genius, they at once constructed a theory which was adequate for their immediate purposes.

Of course, substance and quality, as well as simple location, are the most natural ideas for the human mind. It is the way in which we think of things, and without these ways of thinking we could not get our ideas straight for daily use. There is no doubt about this. The only question is: how concretely are we thinking when we consider nature under these conceptions? My point will be, that we are presenting ourselves with simplified editions of immediate matters of fact. When we examine the primary elements of these simplified editions, we shall find that they are in truth only to be justified as being elaborate logical constructions of a high degree of abstraction. Of course, as a point of individual psychology, we get at the ideas by the rough and ready method of suppressing what appear to be irrelevant details. But when we attempt to justify this suppression of irrelevance, we find that, though there are entities left corresponding to the entities we talk about, yet these entities are of a high degree of abstraction.

Thus I hold that substance and quality afford another

instance of the fallacy of misplaced concreteness. Let us consider how the notions of substance and quality arise. We observe an object as an entity with certain characteristics. Furthermore, each individual entity is apprehended through its characteristics. For example, we observe a body; there is something about it which we note. Perhaps, it is hard, and blue, and round, and noisy. We observe something which possesses these qualities: apart from these qualities we do not observe anything at all. Accordingly, the entity is the substratum, or substance, of which we predicate qualities. Some of the qualities are essential, so that apart from them the entity would not be itself; while other qualities are accidental and changeable. In respect to material bodies, the qualities of having a quantitative mass, and of simple location somewhere, were held by John Locke at the close of the seventeenth century to be essential qualities. Of course, the location was changeable, and the unchangeability of mass was merely an experimental fact for some extremists.

So far, so good. But when we pass to blueness and noisiness a new situation has to be faced. In the first place, the body may not be always blue, or noisy. We have already allowed for this by our theory of accidental qualities, which for the moment we may accept as adequate. But in the second place, the seventeenth century exposed a real difficulty. The great physicists elaborated transmission theories of light and sound, based upon their materialistic views of nature. There were two hypotheses as to light: either it was transmitted by the vibratory waves of a materialistic ether, or – according to Newton – it was transmitted by the motion of incredibly small corpuscles of some subtle matter. We all know that the wave theory of Huyghens held the field during the nineteenth century, and at present physicists are endeavouring to explain some obscure circumstances attending radiation by a combination of both theories. But whatever theory you choose, there is no light or colour as a fact in external nature. There is merely motion of material. Again, when the light enters your eyes and falls on the retina, there is merely motion of material. Then your nerves are affected and your brain is affected, and again this is merely motion of material. The same line of argument holds for sound, substituting waves in the air for waves in the ether, and ears for eyes.

We then ask in what sense are blueness and noisiness qualities of the body. By analogous reasoning, we also ask in what sense is its scent a quality of the rose.

Galileo considered this question, and at once pointed out that, apart from eyes, ears, or noses, there would be no colours, sounds or smells. Descartes and Locke elaborated a theory of primary and secondary qualities. For example, Descartes in his sixth *Meditation* says[1] : 'And indeed, as I perceive different sorts of colours, sounds, odours, tastes, heat, hardness, etc., I safely conclude that there are in the bodies from which the diverse perceptions of the senses proceed, certain varieties corresponding to them, although, perhaps, not in reality like them . . .'

Also in his *Principles of Philosophy*, he says: 'That by our senses we know nothing of external objects beyond their figure [or situation], magnitude, and motion.'

Locke, writing with a knowledge of Newtonian dynamics, places mass among the primary qualities of bodies. In short, he elaborates a theory of primary and secondary qualities in accordance with the state of physical science at the close of the seventeenth century. The primary qualities are the essential qualities of substances whose spatio-temporal relationships constitute nature. The orderliness of these relationships constitutes the order of nature. The occurrences of nature are in some way apprehended by minds, which are associated with living bodies. Primarily, the mental apprehension is aroused by the occurrences in certain parts of the correlated body, the occurrences in the brain, for instance. But the mind in apprehending also experiences sensations which, properly speaking, are qualities of the

[1] Translation by Professor John Veitch.

mind alone. These sensations are projected by the mind so as to clothe appropriate bodies in external nature. Thus the bodies are perceived as with qualities which in reality do not belong to them, qualities which in fact are purely the offspring of the mind. Thus nature gets credit which should in truth be reserved for ourselves: the rose for its scent: the nightingale for his song: and the sun for his radiance. The poets are entirely mistaken. They should address their lyrics to themselves, and should turn them into odes of self-congratulation on the excellency of the human mind. Nature is a dull affair, soundless, scentless, colourless; merely the hurrying of material, endlessly, meaninglessly.

However you disguise it, this is the practical outcome of the characteristic scientific philosophy which closed the seventeenth century.

In the first place, we must note its astounding efficiency as a system of concepts for the organization of scientific research. In this respect, it is fully worthy of the genius of the century which produced it. It has held its own as the guiding principle of scientific studies ever since. It is still reigning. Every university in the world organizes itself in accordance with it. No alternative system of organizing the pursuit of scientific truth has been suggested. It is not only reigning, but it is without a rival.

And yet – it is quite unbelievable. This conception of the universe is surely framed in terms of high abstractions, and the paradox only arises because we have mistaken our abstraction for concrete realities.

No picture, however generalized, of the achievements of scientific thought in this century can omit the advance in mathematics. Here as elsewhere the genius of the epoch made itself evident. Three great Frenchmen, Descartes, Desargues, Pascal, initiated the modern period in geometry. Another Frenchman, Fermat, laid the foundations of modern analysis, and all but perfected the methods of the differential calculus. Newton and Leibniz, between them, actually did create the differential calculus as a practical method of mathematical reasoning. When the century ended, mathematics as an instrument for application to physical problems was well established in something of its modern proficiency. Modern pure mathematics, if we except geometry, was in its infancy, and had given no signs of the astonishing growth it was to make in the nineteenth century. But the mathematical physicist had appeared, bringing with him the type of mind which was to rule the scientific world in the next century. It was to be the age of 'victorious analysis'.

The seventeenth century had finally produced a scheme of scientific thought framed by mathematicians, for the use of mathematicians. The great characteristic of the mathematical mind is its capacity for dealing with abstractions; and for eliciting from them clear-cut demonstrative trains of reasoning, entirely satisfactory so long as it is those abstractions which you want to think about. The enormous success of the scientific abstractions, yielding on the one hand matter with its simple location in space and time, on the other hand mind, perceiving, suffering, reasoning, but not interfering, has foisted on to philosophy the task of accepting them as the most concrete rendering of fact.

Thereby, modern philosophy has been ruined. It has oscillated in a complex manner between three extremes. There are the dualists, who accept matter and mind as on equal basis, and the two varieties of monists, those who put mind inside matter, and those who put matter inside mind. But this juggling with abstractions can never overcome the inherent confusion introduced by the ascription of misplaced concreteness to the scientific scheme of the seventeenth century.

CHAPTER IV

THE EIGHTEENTH CENTURY

In so far as the intellectual climates of different epochs can be contrasted, the eighteenth century in Europe was the complete antithesis to the Middle Ages. The contrast is symbolized by the difference between the cathedral of Chartres and the Parisian salons, where D'Alembert conversed with Voltaire. The Middle Ages were haunted with the desire to rationalize the infinite : the men of the eighteenth century rationalized the social life of modern communities, and based their sociological theories on an appeal to the facts of nature. The earlier period was the age of faith, based upon reason. In the later period, they let sleeping dogs lie : it was the age of reason, based upon faith. To illustrate my meaning : St Anselm would have been distressed if he had failed to find a convincing argument for the existence of God, and on this argument he based his edifice of faith, whereas Hume based his *Dissertation on the Natural History of Religion* upon his faith in the order of nature. In comparing these epochs it is well to remember that reason can err, and that faith may be misplaced.

In my previous lecture I traced the evolution, during the seventeenth century, of the scheme of scientific ideas which has dominated thought ever since. It involves a fundamental duality, with material on the one hand, and on the other hand mind. In between there lie the concepts of life, organism, function, instantaneous reality, interaction, order of nature, which collectively form the Achilles heel of the whole system.

I also express my conviction that if we desired to obtain a more fundamental expression of the concrete character of natural fact, the element in this scheme which we should first criticize is the concept of simple location. In view therefore of the importance which this idea will assume in these lectures, I will repeat the meaning which I have attached to this phrase. To say that a bit of matter has simple location means that, in expressing its spatio-temporal relations, it is adequate to state that it is where it is, in a definite finite region of space, and throughout a definite finite duration of time, apart from any essential reference of the relations of that bit of matter to other regions of space and to other durations of time. Again, this concept of simple location is independent of the controversy between the absolutist and the relativist views of space or of time. So long as any theory of space, or of time, can give a meaning, either absolute or relative, to the idea of a definite region of space, and of a definite duration of time, the idea of simple location has a perfectly definite meaning. This idea is the very foundation of the seventeenth-century scheme of nature. Apart from it, the scheme is incapable of expression. I shall argue that among the primary elements of nature as apprehended in our immediate experience, there is no element whatever which possesses this character of simple location. It does not follow, however, that the science of the seventeenth century was simply wrong. I hold that by a process of constructive abstraction we can arrive at abstractions which are the simply located bits of material, and at other abstractions which are the minds included in the scientific scheme. Accordingly, the real error is an example of what I have termed: the fallacy of misplaced concreteness.

The advantage of confining attention to a definite group of abstractions, is that you confine your thoughts to clear-cut definite things, with clear-cut definite relations. Accordingly, if you have a logical head, you can deduce a variety of conclusions respecting the relationships between these abstract entities. Furthermore, if the abstractions are well-founded, that is to say, if they do not abstract from everything that is important in experience, the scientific thought which confines itself to these abstractions will arrive at a variety of important truths relating to our experience of nature. We all know those clear-cut trenchant intellects, immovably encased in a hard shell of abstractions. They hold you to their abstractions by the sheer grip of personality.

The disadvantage of exclusive attention to a group of abstractions, however well-founded, is that, by the nature of the case, you have abstracted from the remainder of things. In so far as the excluded things are important in your experience, your modes of thought are not fitted to deal with them. You cannot think without abstractions; accordingly, it is of the utmost importance to be vigilant in critically revising your modes of abstraction. It is here that philosophy finds its niche as essential to the healthy progress of society. It is the critic of abstractions. A civilization which cannot burst through its current abstractions is doomed to sterility after a very limited period of progress. An active school of philosophy is quite as important for the locomotion of ideas, as is an active school of railway engineers for the locomotion of fuel.

Sometimes it happens that the service rendered by philosophy is entirely obscured by the astonishing success of a scheme of abstractions in expressing the dominant interests of an epoch. This is exactly what happened during the eighteenth century. *Les philosophes* were not philosophers. They were men of genius, clear-headed and acute, who applied the seventeenth-century group of scientific abstractions to the analysis of the unbounded universe. Their triumph, in respect to the circle of ideas mainly interesting to their contemporaries, was overwhelming; whatever did not fit into their scheme was ignored, derided, disbelieved. Their hatred of Gothic architecture symbolizes their lack of sympathy with dim perspectives. It was the age of reason, healthy, manly, upstanding reason; but, of one-eyed reason, deficient in its vision of depth. We cannot overrate the debt of gratitude which we owe to these men. For a thousand years Europe had been a prey to intolerant, intolerable visionaries. The common sense of the eighteenth century, its grasp of the obvious facts of human suffering, and of the obvious demands of human nature, acted on the world like a bath of moral cleansing. Voltaire must have the credit, that he hated injustice, he hated cruelty, he hated senseless repression, and he hated hocus-pocus. Furthermore, when he saw them, he knew them. In these supreme virtues, he was typical of his century, on its better side. But if men cannot live on bread alone, still less can they do so on disinfectants. The age had its limitations; yet we cannot understand the passion with which some of its main positions are still defended, especially in the schools of science, unless we do full justice to its positive achievements. The seventeenth-century scheme of concepts was proving a perfect instrument for research.

This triumph of materialism was chiefly in the sciences of rational dynamics, physics, and chemistry. So far as dynamics and physics were concerned, progress was in the form of direct developments of the main ideas of the previous epoch. Nothing fundamentally new was introduced, but there was an immense detailed development. Special cases were unravelled. It was as though the very Heavens were being opened, on a set plan. In the second half of the century, Lavoisier practically founded chemistry on its present basis. He introduced into it the principle that no material is lost or gained in any chemical transformations. This was the last success of materialistic thought, which has not ultimately proved to be double-edged. Chemical science now only waited for the atomic theory, in the next century.

In this century the notion of the mechanical explanation of all the processes of nature finally hardened into a dogma of science. The notion won through on its merits by reason of an almost miraculous series of triumphs achieved by the mathematical physicists, culminating in the *Mécanique Analytique* of Lagrange, which was published in 1787. Newton's *Principia* was published in 1687, so that exactly one hundred years separates the two great books. This century contains the first period of mathematical physics of the modern type. The publication of Clerk Maxwell's *Electricity and Magnetism* in 1873 marks the close of the second period. Each of these three books introduces new horizons of thought affecting everything which comes after them.

In considering the various topics to which mankind has bent its systematic thought, it is impossible not to be struck with the unequal distribution of ability among the different fields. In almost all subjects there are a few outstanding names. For it requires genius to create a subject as a distinct topic for thought. But in the case of many topics, after a good beginning very relevant to its immediate occasion, the subsequent development appears as a weak series of flounderings, so that the whole subject gradually loses its grip on the evolution of thought. It was far otherwise with mathematical physics. The more you study this subject, the more you will find yourself astonished by the almost incredible triumphs of intellect which it exhibits. The great mathematical physicists of the eighteenth and first few years of the nineteenth century, most of them French, are a case in point: Maupertuis, Clairaut, D'Alembert, Lagrange, Laplace, Fourier, Carnot, form a series of names, such that each recalls to mind some achievement of the first rank. When Carlyle, as the mouthpiece of the subsequent romantic age, scoffingly terms the period the age of victorious analysis, and mocks at Maupertuis as a 'sublimish gentleman in a white periwig', he only exhibits the narrow side of the Romanticists who he is then voicing.

It is impossible to explain intelligently, in a short time and without technicalities, the details of the progress made by this school. I will, however, endeavour to explain the main point of a joint achievement of Maupertuis and Lagrange. Their results, in conjunction with some subsequent mathematical methods due to two great German mathematicians of the first half of the nineteenth century, Gauss and Riemann, have recently proved themselves to be the preparatory work necessary for the new ideas which Herz and Einstein have introduced into mathematical physics. Also they inspired some of the best ideas in Clerk Maxwell's treatise, already mentioned in this lecture.

They aimed at discovering something more fundamental and more general than Newton's laws of motion which were discussed in the previous lecture. They wanted to find some wider ideas, and in the case of Lagrange some more general means of mathematical exposition. It was an ambitious enterprise, and they were completely successful. Maupertuis lived in the first half of the eighteenth century, and Lagrange's active life lay in its second half. We find in Maupertuis a tinge of the theologic age which preceded his birth. He started with the idea that the whole path of a material particle between any limits of time must achieve some perfection worthy of the providence of God. There are two points of interest in this motive principle. In the first place, it illustrates the thesis which I was urging in my first lecture that the way in which the medieval Church had impressed on Europe the notion of the detailed providence of a rational personal God was one of the factors by which the trust in the order of nature had been generated. In the second place, though we are now all convinced that such modes of thought are of no direct use in detailed scientific inquiry, Maupertuis' success in this particular case shows that almost any idea which jogs you out of your current abstractions may be better than nothing. In the present case what the idea in question did for Maupertuis was to lead him to inquire what general property of the path as a whole could be deduced from Newton's laws of motion. Undoubtedly this was a very sensible procedure whatever one's theological notions. Also his general idea led him to conceive that the property found would be a quantitative sum, such that any slight deviation from the path would increase it. In this supposition he was generalizing Newton's first law of motion. For an isolated particle takes the shortest route with uniform velocity. So Maupertuis conjectured that a particle travelling through a field of force would realize the least possible amount of some quantity. He discovered such a quantity and called it the integral action between the time limits considered. In modern phraseology it is the sum through successive small lapses of time of the difference between the kinetic and potential energies of the particle at each successive instant. This action, therefore, has to do with the interchange between the energy arising from motion and the energy arising from position. Maupertuis had discovered the famous theorem of least action. Maupertuis was not quite of the first rank in comparison with such a man as Lagrange. In his hands and in those of his immediate successors, his principle did not assume any dominating importance. Lagrange put the same question on a wider basis so as to make its answer relevant to actual procedure in the development of dynamics. His principle of virtual work as applied to systems in motion is in effect Maupertuis' principle conceived as applying at each instant of the path of the system. But Lagrange saw further than Maupertuis. He grasped that he had gained a method of stating dynamical truths in a way which is perfectly indifferent to the particular methods of measurement employed in fixing the positions of the various parts of the system. Accordingly, he went on to deduce equations of motion which are equally applicable whatever quantitative measurements have been made, provided that they are adequate to fix positions. The beauty and almost divine simplicity of these equations is such that these formulae are worthy to rank with those mysterious symbols which in ancient times were held directly to indicate the supreme reason at the base of all things. Later Herz – inventor of electromagnetic waves – based mechanics on the idea of every particle traversing the shortest path open to it under the circumstances constraining its motion; and finally Einstein, by the use of the geometrical theories of Gauss and Riemann, showed that these circumstances could be construed as being inherent in the character of space-time itself. Such, in barest outline, is the story of dynamics from Galileo to Einstein.

Meanwhile Galvani and Volta lived and made their electric discoveries; and the biological sciences slowly gathered their material, but still waited for dominating ideas. Psychology, also, was beginning to disengage itself from its dependence on general philosophy. This independent growth of psychology was the ultimate result of its invocation by John Locke as a critic of metaphysical licence. All the sciences dealing with life were still in an elementary observational stage, in which classification and direct description were dominant. So far the scheme of abstractions was adequate to the occasion.

In the realm of practice, the age which produced enlightened rulers, such as the Emperor Joseph of the House of Hapsburg, Frederick the Great, Walpole, the great Lord Chatham, George Washington, cannot be said to have failed. Especially when to these rulers, it adds the invention of parliamentary cabinet government in England, of federal presidential government in the United States, and of the humanitarian principles of the French Revolution. Also in technology it produced the steam-engine, and thereby ushered in a new era of civilization. Undoubtedly, as a practical age the eighteenth century was a success. If you had asked one of the wisest and most typical of its ancestors, who just saw its commencement, I mean John Locke, what he expected from it he would hardly have pitched his hopes higher than its actual achievements.

In developing a criticism of the scientific scheme of the eighteenth century, I must first give my main reason for ignoring nineteenth-century idealism – I am speaking of the philosophic idealism which finds the ultimate meaning of reality in mentality that is fully cognitive. The idealistic school, as hitherto developed, has been too much divorced from the scientific outlook. It has swallowed the scientific scheme in its entirety as being the only rendering of the facts of nature, and has then explained it as being an idea in the ultimate mentality. In the case of absolute idealism, the world of nature is just one of the ideas, somehow differentiating the unity of the Absolute : in the case of pluralistic idealism involving monadic mentalities, this world is the greatest common measure of the various ideas which differentiate the various mental unities of the various monads. But, however you take it, these idealistic schools have conspicuously failed to connect, in any organic fashion, the fact of nature with their idealist philosophies. So far as concerns what will be said in these lectures, your ultimate outlook may be realistic or idealistic. My point is that a further stage of provisional realism is required in which the scientific scheme is recast, and founded upon the ultimate concept of *organism*.

In outline, my procedure is to start from the analysis of the status of space and of time, or in modern phraseology, the status of space-time. There are two characters of either. Things are separated by space, and are separated by time : but they are also together in space, and together in time, even if they be not contemporaneous. I will call these characters the 'separative' and the 'prehensive' characters of space-time. There is yet a third character of space-time. Everything which is in space receives a definite limitation of some sort, so that in a sense it has just that shape which it does have and no other, also in some sense it is just in this place and in no other. Analogously for time, a thing endures during a certain period, and through no other period. I will call this the 'modal' character of space-time. It is evident that the modal character taken by itself gives rise to the idea of simple location. But it must be conjoined with the separative and prehensive characters.

For simplicity of thought, I will first speak of space only, and will afterwards extend the same treatment to time.

The volume is the most concrete element of space. But the separative character of space, analyses a volume into sub-volumes, and so on indefinitely. Accordingly, taking the separative character in isolation, we should infer that a volume is a mere multiplicity of non-voluminous elements, of points in fact. But it is the unity of volume which is the ultimate fact of experience, for example, the voluminous space of this hall. This hall as a mere multiplicity of points is a construction of the logical imagination.

Accordingly, the prime fact is the prehensive unity of volume, and this unity is mitigated or limited by the separated unities of the innumerable contained parts. We have a prehensive unity, which is yet held apart as an aggregate of contained parts. But the prehensive unity of the volume is not the unity of a mere logical aggregate of parts. The parts form an ordered aggregate, in the sense that each part is something from the standpoint of every other part, and also from the same standpoint every other part is something in relation to it. Thus if *A* and *B* and *C* are volumes of space, *B* has an aspect from the standpoint of *A*, and so has *C*, and so has the relationship of *B* and *C*. This aspect of *B* from *A* is of the essence of *A*. The volumes of space have no independent existence. They are only entities as within the totality; you cannot extract them from their environment without destruction of their very essence. Accordingly, I will say that the aspect of *B* from *A* is the *mode* in which *B* enters into the composition of *A*. This is the modal character of space, that the prehensive unity of *A* is the prehension into unity of the aspects of all other volumes from the standpoint of *A*. The shape of a volume is the formula from which the totality of its aspects can be derived. Thus the shape of a volume is more abstract than its aspects. It is evident that I can use Leibniz's language, and say that every volume mirrors in itself every other volume in space.

Exactly analogous considerations hold with respect to durations in time. An instant of time, without duration, is an imaginative logical construction. Also each duration of time mirrors in itself all temporal durations.

But in two ways I have introduced a false simplicity. In the first place, I should have conjoined space and time, and conducted my explanation in respect to four-dimensional regions of space-time. I have nothing to add in the way of explanation. In your minds, substitute such four-dimensional regions for the spatial volumes of the previous explanations.

Secondly, my explanation has involved itself in a vicious circle. For I have made the prehensive unity of the region *A* to consist of the prehensive unification of the modal presences in *A* of other regions. This difficulty arises because space-time cannot in reality be considered as a self-subsistent entity. It is an abstraction, and its explanation requires reference to that from which it has been extracted. Space-time is the specification of certain general characters of events and of their mutual ordering. This recurrence to concrete fact brings me back to the eighteenth century, and indeed to Francis Bacon in the seventeenth century. We have to consider the development in those epochs of the criticism of the reigning scientific scheme.

No epoch is homogeneous; whatever you may have assigned as the dominant note of a considerable period, it will always be possible to produce men, and great men, belonging to the same time, who exhibit themselves as antagonistic to the tone of their age. This is certainly the case with the eighteenth century. For example, the names of John Wesley and of Rousseau must have occurred to you while I was drawing the character of that time. But I do not want to speak of them, or of others. The man, whose ideas I must consider at some length, is Bishop Berkeley. Quite at the commencement of the epoch, he made all the right criticisms, at least in principle. It would be untrue to say that he produced no effect. He was a famous man. The wife of George II was one of the few queens who, in any country, have been clever enough, and wise enough, to patronize learning judiciously; accordingly, Berkeley was made a bishop, in days when bishops in Great Britain were relatively far greater men than they are now. Also, what was more important than his bishopric, Hume studied him, and developed one side of his philosophy in a way which might have disturbed the ghost of the great ecclesiastic. Then Kant studied Hume. So, to say that Berkeley was uninfluential during the century, would certainly be absurd. But all the same, he failed to affect the main stream of scientific thought. It flowed on as if he had never written. Its general success made it impervious to criticism, then and since. The world of science has always remained perfectly satisfied with its peculiar abstractions. They work, and that is sufficient for it.

The point before us is that this scientific field of thought is now, in the twentieth century, too narrow for the concrete facts which are before it for analysis. This is true even in physics, and is more especially urgent in the biological sciences. Thus, in order to understand the difficulties of modern scientific thought and also its reactions on the modern world, we should have in our minds some conception of a wider field of abstraction, a more concrete analysis, which shall stand nearer to the complete concreteness of our intuitive experience. Such an analysis should find in itself a niche for the concepts of matter and spirit, as abstractions in terms of which much of our physical experience can be interpreted. It is in the search for this wider basis for scientific thought that Berkeley is so important. He launched his criticism shortly after the schools of Newton and Locke had completed their work, and laid his finger exactly on the weak spots which they had left. I do not propose to consider either the subjective idealism which has been derived from him, or the schools of development which trace their descent from Hume and Kant respectively. My point will be that – whatever the final metaphysics you may adopt – there is another line of development embedded in Berkeley, pointing to the analysis which we are in search of. Berkeley overlooked it, partly by reason of the over-intellectualism of philosophers, and partly by his haste to have recourse to an idealism with its objectivity grounded in the mind of God. You will remember that I have already stated that the key of the problem lies in the notion of simple location. Berkeley, in effect, criticizes this notion. He also raises the question, What do we mean by things being realized in the world of nature?

In Sections 23 and 24 of his *Principles of Human Knowledge*, Berkeley gives his answer to this latter question. I will quote some detached sentences from those Sections:

> 23. But, say you, surely there is nothing easier than for me to imagine trees, for instance, in a park, or books existing in a closet, and nobody by to perceive them. I answer, you may so, there is no difficulty in it; but what is all this, I beseech you, more than framing in your mind certain ideas which you call books and trees, and at the same time omitting to frame the idea of any one that may perceive them? . . .
>
> When we do our utmost to conceive the existence of external bodies, we are all the while only contemplating our own ideas. But the mind *taking no notice of itself*, is deluded to think it can and does conceive bodies existing unthought of or without the mind, though at the same time they are apprehended by or exist in itself . . .
>
> 24. It is very obvious, upon the least inquiry into our thoughts, to know whether it be possible for us to understand what is meant by the *absolute existence of sensible objects in themselves, or without the mind*. To me it is evident those words mark out either a direct contradiction, or else nothing at all . . .

Again there is a very remarkable passage in Section 10, of the fourth Dialogue of Berkeley's *Alciphron*. I have already quoted it, at greater length, in my *Principles of Natural Knowledge*:

> *Euphranor*. Tell me, Alciphron, can you discern the doors, window and battlements of that same castle?
>
> *Alciphron*. I cannot. At this distance it seems only a small round tower.
>
> *Euph*. But I, who have been at it, know that it is no

small round tower, but a large square building with battlements and turrets, which it seems you do not see.

Alc. What will you infer from thence?

Euph. I would infer that the very object which you strictly and properly perceive by sight is not that thing which is several miles distant.

Alc. Why so?

Euph. Because a little round object is one thing, and a great square object is another. It is not so? . . .

Some analogous examples concerning a planet and a cloud are then cited in the dialogue, and this passage finally concludes with:

> *Euphranor.* Is it not plain, therefore, that neither the castle, the planet, nor the cloud, *which you see here*, are those real ones which you suppose exist at a distance?

It is made explicit to the first passage, already quoted, that Berkeley himself adopts an extreme idealistic interpretation. For him mind is the only absolute reality, and the unity of nature is the unity of ideas in the mind of God. Personally, I think that Berkeley's solution of the metaphysical problem raises difficulties not less than those which he points out as arising from a realistic interpretation of the scientific scheme. There is, however, another possible line of thought, which enables us to adopt anyhow an attitude of provisional realism, and to widen the scientific scheme in a way which is useful for science itself.

I recur to the passage from Francis Bacon's *Natural History*, already quoted in the previous lecture:

> It is certain that all bodies whatsoever, though they have no sense, yet they have perception: . . . and whether the body be alterant or altered, evermore a perception precedeth operation; for else all bodies would be alike one to another . . .

Also in the previous lecture I construed perception (as used by Bacon) as meaning taking account of the essential character of the thing perceived, and I construed sense as meaning cognition. We certainly do take account of things of which at the time we have no explicit cognition. We can even have a cognitive memory of the taking account, without having had a contemporaneous cognition. Also, as Bacon points out by his statement, '. . . for else all bodies would be alike one to another,' it is evidently some element of the essential character which we take account of, namely something on which diversity is founded and not mere bare logical diversity.

The word 'perceive' is, in our common usage, shot through and through with the notion of cognitive apprehension. So is the word 'apprehension', even with the adjective cognitive omitted. I will use the word 'prehension' for uncognitive apprehension : by this I mean apprehension which may or may not be cognitive. Now take Euphranor's last remark :

'Is it not plain, therefore, that neither the castle, the planet, nor the cloud, *which you see here*, are those real ones which you suppose exist at a distance?' Accordingly, there is a prehension, *here* in this place, of things which have a reference to *other* places.

Now go back to Berkeley's sentences, quoted from his *Principles of Human Knowledge*. He contends that what constitutes the realization of natural entities is the being perceived within the unity of mind.

We can substitute the concept, that the realization is a gathering of things into the unity of a prehension; and that what is thereby realized is the prehension, and not the things. This unity of a prehension defines itself as a *here* and a *now*, and the things so gathered into the grasped unity have essential reference to other places and other times. For Berkeley's mind, I substitute a process of prehensive unification. In order to make intelligible this concept of the progressive realization of natural occurrences, considerable expansion is required, and confrontation with its actual implications in terms of concrete experience. This will be the task of the subsequent lectures. In the first place, note the idea of simple location has gone. The things which are grasped into a realized unity, here and now, are not the castle, the cloud, and the planet simply in themselves; but they are the castle, the cloud, and the planet from the standpoint, in space and time, of the prehensive unification. In other words, it is the perspective of the castle over there from the standpoint of the unification here. It is, therefore, aspects of the castle, the cloud, and the planet which are grasped into unity here. You will remember that the idea of perspectives is quite familiar in philosophy. It was introduced by Leibniz, in the notion of his monads mirroring perspectives of the universe. I am using the same notion, only I am toning down his monads into the unified events in space and time. In some ways, there is a greater analogy with Spinoza's modes; that is why I use the terms 'mode' and 'modal'. In the analogy with Spinoza, his one substance is for me the one underlying activity of realization individualizing itself in an interlocked plurality of modes. Thus, concrete fact is process. Its primary analysis is into underlying activity of prehension, and into realized prehensive events. Each event is an individual matter of fact issuing from an individualization of the substrate activity. But individualization does not mean substantial independence.

An entity of which we become aware in sense perception is the terminus of our act of perception. I will call such an entity, a 'sense-object'. For example, a green of a definite shade is a sense-object; so is a sound of definite quality and pitch; and so is a definite scent; and a definite quality of touch. The way in which such an entity is related to space during a definite lapse of time is complex. I will say that a sense-object has 'ingression' into space-time. The cognitive perception of a sense-object is the awareness of the prehensive unification (into a standpoint *A*) of various modes of various sense-objects, including the sense-object in question. The standpoint *A* is, of course, a region of space-time; that is to say, it is a volume of space through a duration of time. But as one entity, this standpoint is a unit of realized experience. A mode of a sense-object at *A* (as abstracted from the sense-object whose relationship to *A* the mode is conditioning) is the aspect from *A* of some other region *B*. Thus the sense-object is present in *A* with the mode of location in *B*. Thus if green be the sense-object in question, green is not simply at *A* where it is being perceived, nor is it simply at *B* where it is perceived as located; but it is present at *A* with the mode of location in *B*. There is no particular mystery about this. You have only got to look into a mirror and to see the image in it of some green leaves behind your back. For you at *A* there will be green; but not green simply at *A* where you are. The green at *A* will be green with the mode of having location at the image of the leaf behind the mirror. Then turn round and look at the leaf. You are now perceiving the green in the same way as you did before, except that now the green has the mode of being located in the actual leaf. I am merely describing what we do perceive: we are aware of green as being one element in a prehensive unification of sense-objects; each sense-object, and among them green, having its particular mode, which is expressible as location elsewhere. There are various types of modal location. For example, sound is voluminous: it fills a hall, and so sometimes does diffused colour. But the modal location of a colour may be that of being the remote boundary of a volume, as for example the colours on the walls of a room. Thus primarily space-time is the locus of the modal ingression of sense-objects. This is the reason why space and time (if for simplicity we disjoin them) are given in their entireties. For each volume of space, or each lapse of time, includes in its essence aspects of all volumes of space, or of all lapses of time. The difficulties of philosophy in respect to space and time are founded on the error of considering them as primarily the loci of simple locations. Perception is simply the cognition of prehensive unification; or more shortly, perception is cognition of prehension. The actual world is a manifold of prehensions; and a 'prehension' is a 'prehensive occasion'; and a prehensive occasion is the most concrete finite entity, conceived as what it is in itself and for itself, and not as from its aspect in the essence of another such occasion. Prehensive unification might be said to have simple location in its volume *A*. But this would be a mere tautology. For space and time are simply abstractions from the totality of prehensive unifications as mutually patterned in each other. Thus a prehension has simple location at the volume *A* in the same way as that in which a man's face fits on to the smile which spreads over it. There is, so far as we have gone, more sense in saying that an act of perception has simple location; for it may be conceived as being simply at the cognized prehension.

There are more entities involved in nature than the mere sense-objects, so far considered. But, allowing for the necessity of revision consequent on a more complete point of view, we can frame our answer to Berkeley's question as to the character of the reality to be assigned to nature. He states it to be the reality of ideas in mind. A complete metaphysic which has attained to some notion of mind, and to some notion of ideas, may perhaps ultimately adopt that view. It is unnecessary for the purpose of these lectures to ask such a fundamental question. We can be content with a provisional realism in which nature is conceived as a complex of prehensive unifications. Space and time exhibit the general scheme of interlocked relations of these prehensions. You cannot tear any one of them out of its context. Yet each one of them within its context has all the reality that attaches to the whole complex. Conversely, the totality has the same reality as each prehension; for each prehension unifies the modalities to be ascribed, from its standpoint, to every part of the whole. A prehension is a process of unifying. Accordingly, nature is a process of expansive development, necessarily transitional from prehension to prehension. What is achieved is thereby passed beyond, but it is also retained as having aspects of itself present to prehensions which lie beyond it.

Thus nature is a structure of evolving processes. The reality is the process. It is nonsense to ask if the colour red is real. The colour red is ingredient in the process of realization. The realities of nature are the prehensions in nature, that is to say, the events in nature.

Now that we have cleared space and time from the taint of simple location, we may partially abandon the awkward term prehension. This term was introduced to signify the essential unity of an event, namely, the event as one entity, and not as a mere assemblage of parts or of ingredients. It is necessary to understand that space-time is nothing else than a system of pulling together of assemblages into unities. But the word event just means one of these spatio-temporal unities. Accordingly, it may be used instead of the term 'prehension' as meaning the thing prehended.

An event has contemporaries. This means that an event mirrors within itself the modes of its contemporaries as a display of immediate achievement. An event has a past. This means that an event mirrors within itself the modes of its predecessors, as memories which are fused into its own content. An event has a future. This means that an event mirrors within itself such aspects as the future throws back on to the present, or, in other words, as the present has determined concerning the future. Thus an event has anticipation :

> The prophetic soul
> Of the wide world dreaming on things to come. [cvii]

These conclusions are essential for any form of realism. For there is in the world for our cognizance, memory of the past, immediacy of realization, and indication of things to come.

In this sketch of an analysis more concrete than that of the scientific scheme of thought, I have started from our own psychological field, as it stands for our cognition. I take it for what it claims to be: the self-knowledge of our bodily event. I mean the total event, and not the inspection of the details of the body. This self-knowledge discloses a prehensive unification of modal presences of entities beyond itself. I generalize by the use of the principle that this total bodily event is on the same level as all other events, except for an unusual complexity and stability of inherent pattern. The strength of the theory of materialistic mechanism has been the demand, that no arbitrary breaks be introduced into nature, to eke out the collapse of an explanation. I accept this principle. But if you start from the immediate facts of our psychological experience, as surely an empiricist should begin, you are at once led to the organic conception of nature of which the description has been commenced in this lecture.

It is the defect of the eighteenth-century scientific scheme that it provides none of the elements which compose the immediate psychological experiences of mankind. Nor does it provide any elementary trace of the organic unity of a whole, from which the organic unities of electrons, protons, molecules, and living bodies can emerge. According to that scheme, there is no reason in the nature of things why portions of material should have any physical relations to each other. Let us grant that we cannot hope to be able to discern the laws of nature to be necessary. But we can hope to see that it is necessary that there should be an order of nature. The concept of the order of nature is bound up with the concept of nature as the locus of organisms in process of development.

NOTE. In connection with the latter portion of this chapter a sentence from Descates' 'Reply to Objections . . . against the Meditations' is interesting: 'Hence the idea of the sun will be the sun itself existing in the mind, not indeed formally, as it exists in the sky, but objectively, i.e., in the way in which objects are wont to exist in the mind; and this mode of being is truly much less perfect than that in which things exist outside in the mind, but it is not on that account mere nothing, as I have already said.' (Reply to Objections I, Translation by Haldane and Ross, vol. ii, p. 10.) I find difficulty in reconciling this theory of ideas (with which I agree) with other parts of the Cartesian philosophy.

CHAPTER V

THE ROMANTIC REACTION

My last lecture described the influence upon the eighteenth century of the narrow and efficient scheme of scientific concepts which it had inherited from its predecessor. That scheme was the product of a mentality which found the Augustinian theology extremely congenial. The Protestant Calvinism and the Catholic Jansenism exhibited man as helpless to co-operate with irresistible grace: the contemporary scheme of science exhibited man as helpless to co-operate with the irresistible mechanism of nature. The mechanism of God and the mechanism of matter were the monstrous issues of limited metaphysics and clear logical intellect. Also the seventeenth century had genius, and cleared the world of muddled thought. The eighteenth century continued the work of clearance, with ruthless efficiency. The scientific scheme has lasted longer than the theological scheme. Mankind soon lost interest in irresistible grace; but it quickly appreciated the competent engineering which was due to science. Also in the first quarter of the eighteenth century, George Berkeley launched his philosophical criticism against the whole basis of the system. He failed to disturb the dominant current of thought. In my last lecture I developed a parallel line of argument, which would lead to a system of thought basing nature upon the concept of organism, and not upon the concept of matter. In the present lecture, I propose in the first place to consider how the concrete educated thought of men has viewed this opposition of mechanism and organism. It is in literature that the concrete outlook of humanity receives its expression. Accordingly it is to literature that we must look, particularly in its more concrete forms, namely in poetry and in drama, if we hope to discover the inward thoughts of a generation.

We quickly find that the Western peoples exhibit on a colossal scale a peculiarity which is popularly supposed to be more especially characteristic of the Chinese. Surprise is often expressed that a Chinese can be of two religions, a Confucian for some occasions and a Buddhist for other occasions. Whether this is true of China I do not know; nor do I know whether, if true, these two attitudes are really inconsistent. But there can be no doubt that an analogous fact is true of the West, and that the two attitudes involved are inconsistent. A scientific realism, based on mechanism, is conjoined with an unwavering belief in the world of men and of the higher animals as being composed of self-determining organisms. This radical inconsistency at the basis of modern thought accounts for much that is half-hearted and wavering in our civilization. It would be going too far to say that it distracts thought. It enfeebles it, by reason of the inconsistency lurking in the background. After all, the men of the Middle Ages were in pursuit of an excellency of which we have nearly forgotten the existence. They set before themselves the ideal of the attainment of a harmony of the understanding. We are content with superficial orderings from diverse arbitrary starting points. For instance, the enterprises produced by the individualistic energy of the European peoples presuppose physical actions directed to final causes. But the science which is employed in their development is based on a philosophy which asserts that physical causation is supreme, and which disjoins the physical cause from the final end. It is not popular to dwell on the absolute contradiction here involved. It is the fact however you gloze it over with phrases. Of course, we find in the eighteenth century Paley's famous argument, that mechanism presupposes a God who is the author of nature. But even before Paley put the argument into its final form, Hume had written the retort, that the

God whom you will find will be the sort of God who makes that mechanism. In other words, that mechanism can, at most, presuppose a mechanic, and not merely *a* mechanic but *its* mechanic. The only way of mitigating mechanism is by the discovery that it is not mechanism.

When we leave apologetic theology, and come to ordinary literature, we find, as we might expect, that the scientific outlook is in general simply ignored. So far as the mass of literature is concerned, science might never have been heard of. Until recently nearly all writers have been soaked in classical and renaissance literature. For the most part, neither philosophy nor science interested them, and their minds were trained to ignore it.

There are exceptions to this sweeping statement; and, even if we confine ourselves to English literature, they concern some of the greatest names; also the indirect influence of science has been considerable.

A side light on this distracting inconsistency in modern thought is obtained by examining some of those great serious poems in English literature, whose general scale gives them a didactic character. The relevant poems are Milton's *Paradise Lost*, Pope's *Essay on Man*, Wordsworth's *Excursion*, Tennyson's *In Memoriam*. Milton, though he is writing after the Restoration, voices the theological aspect of the earlier portion of his century, untouched by the influence of the scientific materialism. Pope's poem represents the effect on popular thought of the intervening sixty years which includes the first period of assured triumph for the scientific movement. Wordsworth in his whole being expresses a conscious reaction against the mentality of the eighteenth century. This mentality means nothing else than the acceptance of the scientific ideas at their full face value. Wordsworth was not bothered by any intellectual antagonism. What moved him was a moral repulsion. He felt that something had been left out, and that what had been left out comprised everything that was most important. Tennyson is the mouthpiece of the attempts of the waning romantic movement in the second quarter of the nineteenth century to come to terms with science. By this time the two elements in modern thought had disclosed their fundamental divergence by their jarring interpretations of the course of nature and the life of man. Tennyson stands in this poem as the perfect example of the distraction which I have already mentioned. There are opposing visions of the world, and both of them command his assent by appeals to ultimate intuitions from which there seems no escape. Tennyson goes to the heart of the difficulty. It is the problem of mechanism which appalls him,

'The stars,' she whispers, 'blindly run.'

This line states starkly the whole philosophic problem implicit in the poem. Each molecule blindly runs. The human body is a collection of molecules. Therefore, the human body blindly runs, and therefore there can be no individual responsibility for the actions of the body. If you once accept that the molecule is definitely determined to be what it is, independently of any determination by reason of the total organism of the body, and if you further admit that the blind run is settled by the general mechanical laws, there can be no escape from this conclusion. But mental experiences are derivative from the actions of the body, including of course its internal behaviour. Accordingly, the sole function of the mind is to have at least some of its experiences settled for it, and to add such others as may be open to it independently of the body's motions, internal and external.

There are then two possible theories as to the mind. You can either deny that it can supply for itself any experiences other than those provided for it by the body, or you can admit them.

If you refuse to admit the additional experiences, then all individual moral responsibility is swept away. If you do admit them, then a human being may be responsible for the actions of his body. The enfeeblement of thought in the modern world is illustrated by the way in which this plain issue is avoided in Tennyson's poem. There is something kept in the background, a skeleton in the cupboard. He touches on almost every religious and scientific problem, but carefully avoids more than a passing allusion to this one.

This very problem was in full debate at the date of the poem. John Stuart Mill was maintaining his doctrine of determinism. In this doctrine volitions are determined by motives, and motives are expressible in terms of antecedent conditions including states of mind as well as states of the body.

It is obvious that this doctrine affords no escape from the dilemma presented by a thoroughgoing mechanism. For if the volition affects the state of the body, then the molecules in the body do not blindly run. If the volition does not affect the state of the body, the mind is still left in its uncomfortable position.

Mill's doctrine is generally accepted, especially among scientists, as though in some way it allowed you to accept the extreme doctrine of materialistic mechanism, and yet mitigated its unbelievable consequences. It does nothing of the sort. Either the bodily molecules blindly run, or they do not. If they do blindly run, the mental states are irrelevant in discussing the bodily actions.

I have stated the arguments concisely, because in truth the issue is a very simple one. Prolonged discussion is merely a source of confusion. The question as to the metaphysical status of molecules does not come in. The statement that they are mere formulae has no bearing on the argument. For presumably the formulae mean something. If they mean nothing, the whole mechanical doctrine is likewise without meaning, and the question drops. But if the formulae mean anything, the argument applies to exactly what they do mean. The traditional way of evading the difficulty – other than the simple way of ignoring it – is to have recourse to some form of what is now termed 'vitalism'. This doctrine is really a compromise. It allows a free run to mechanism throughout the whole of inanimate nature, and holds that the mechanism is partially mitigated within living bodies. I feel that this theory is an unsatisfactory compromise. The gap between living and dead matter is too vague and problematical to bear the weight of such an arbitrary assumption, which involves an essential dualism somewhere.

The doctrine which I am maintaining is that the whole concept of materialism only applies to very abstract entities, the products of logical discernment. The concrete enduring entities are organisms, so that the plan of the whole influences the very characters of the various subordinate organisms which enter into it. In the case of an animal, the mental states enter into the plan of the total organism and thus modify the plans of the successive subordinate organisms until the ultimate smallest organisms, such as electrons, are reached. Thus an electron within a living body is different from an electron outside it, by reason of the plan of the body. The electron blindly runs either within or without the body; but it runs within the body in accordance with its character within the body; that is to say, in accordance with the general plan of the body, and this plan includes the mental state. But the principle of modification is perfectly general throughout nature, and represents no property peculiar to living bodies. In subsequent lectures it will be explained that this doctrine involves the abandonment of the traditional scientific materialism, and the substitution of an alternative doctrine of organism.

I shall not discuss Mill's determinism, as it lies outside the scheme of these lectures. The foregoing discussion has been directed to secure that either determinism or free will shall have some relevance, unhampered by the difficulties introduced by materialistic mechanism, or by the compromise of vitalism. I would term the doctrine of these lectures, the theory of organic mechanism. In this theory, the molecules may blindly run in accordance with the general laws, but the molecules differ in their intrinsic characters according to the general organic plans of the situations in which they find themselves.

The discrepancy between the materialistic mechanism of science and the moral intuitions, which are presupposed in the concrete affairs of life, only gradually assumed its true importance as the centuries advanced. The different tones of the successive epochs to which the poems, already mentioned, belong are curiously reflected in their opening passages. Milton ends his introduction with the prayer,

> That to the height of this great argument
> I may assert eternal Providence,
> And justify the ways of God to men.

To judge from many modern writers on Milton, we might imagine that the *Paradise Lost* and the *Paradise Regained* were written as a series of experiments in blank verse. This was certainly not Milton's view of his work. To 'justify the ways of God to men' was very much his main object. He recurs to the same idea in the *Samson Agonistes*,

> Just are the ways of God
> And justifiable to men.

We note the assured volume of confidence, untroubled by the coming scientific avalanche. The actual date of the publication of the *Paradise Lost* lies just beyond the epoch to which it belongs. It is the swan-song of a passing world of untroubled certitude.

A comparison between Pope's *Essay on Man* and the *Paradise Lost* exhibits the change of tone in English thought in the fifty or sixty years which separate the age of Milton from the age of Pope. Milton addresses his poem to God. Pope's poem is addressed to Lord Bolingbroke,

Awake, my St John! leave all meaner things
To low ambition and the pride of kings.
Let us (since life can little more supply
Than just to look about us and to die)
Expatiate free o'er all this scene of man;
A mighty maze! but not without a plan.

Compare the jaunty assurance of Pope,

A mighty maze! but not without a plan

with Milton's

Just are the ways of God
And justifiable to men.

But the real point to notice is that Pope as well as Milton was untroubled by the great perplexity which haunts the modern world. The clue which Milton followed was to dwell on the ways of God in dealings with man. Two generations later we find Pope equally confident that the enlightened methods of modern science provided a plan adequate as a map of the 'mighty maze'.

Wordsworth's *Excursion* is the next English poem on the same subject. A prose preface tells us that it is a fragment of a larger projected work, described as 'A philosophical poem containing views of Man, Nature, and Society.'

Very characteristically the poem begins with the line,

'Twas summer, and the sun had mounted high.

Thus the romantic reaction started neither with God nor with Lord Bolingbroke, but with nature. We are here witnessing a conscious reaction against the whole tone of the eighteenth century. That century approached nature with the abstract analysis of science, whereas Wordsworth opposes to the scientific abstractions his full concrete experience.

A generation of religious revival and of scientific advance lies between the *Excursion* and Tennyson's *In Memoriam*. The earlier poets had solved the perplexity by ignoring it. That course was not open to Tennyson. Accordingly his poem begins thus:

> Strong Son of God, immortal Love,
> Whom we, that have not seen Thy face,
> By faith, and faith alone, embrace,
> Believing where we cannot prove.

The note of perplexity is struck at once. The nineteenth century has been a perplexed century, in a sense which is not true of any of its predecessors of the modern period. In the earlier times there were opposing camps, bitterly at variance on questions which they deemed fundamental. But, except for a few stragglers, either camp was whole-hearted. The importance of Tennyson's poem lies in the fact that it exactly expressed the character of its period. Each individual was divided against himself. In the earlier times, the deep thinkers were the clear thinkers – Descartes, Spinoza, Locke, Leibniz. They knew exactly what they meant and said it. In the nineteenth century, some of the deeper thinkers among theologians and philosophers were muddled thinkers. Their assent was claimed by incompatible doctrines; and their efforts at reconciliation produced inevitable confusion.

Matthew Arnold, even more than Tennyson, was the poet who expressed this mood of individual distraction which was so characteristic of this century. Compare with *In Memoriam* the closing lines of Arnold's *Dover Beach*:

> And we are here as on a darkling plain
> Swept with confused alarms of struggle and flight,
> Where ignorant armies clash by night.

Cardinal Newman in his *Apologia Pro Vita Sua* mentions it as a peculiarity of Pusey, the great Anglican ecclesiastic, 'He was haunted by no intellectual perplexities.' In this respect Pusey recalls Milton, Pope, Wordsworth, as in contrast with Tennyson, Clough, Matthew Arnold, and Newman himself.

So far as concerns English literature we find, as might be anticipated, the most interesting criticism of the thoughts of science among the leaders of the romantic reaction which accompanied and succeeded the epoch of the French Revolution. In English literature, the deepest thinkers of this school were Coleridge, Wordsworth, and Shelley. Keats is an example of literature untouched by science. We may neglect Coleridge's attempt at an explicit philosophical formulation. It was influential in his own generation; but in these lectures it is my object only to mention those elements of the thought of the past which stand for all time. Even with this limitation, only a selection is possible. For our purposes Coleridge is only important by his influence on Wordsworth. Thus Wordsworth and Shelley remain.

Wordsworth was passionately absorbed in nature. It has been said of Spinoza, that he was drunk with God. It is equally true that Wordsworth was drunk with nature. But he was a thoughtful, well-read man, with philosophical interests, and sane even to the point of prosiness. In addition, he was a genius. He weakens his evidence by his dislike of science. We all remember his scorn of the poor man whom he somewhat hastily accuses of peeping and botanizing on his mother's grave. Passage after passage could be quoted from him, expressing this repulsion. In this respect, his characteristic thought can be summed up in his phrase, 'We murder to dissect.'

In this latter passage, he discloses the intellectual basis of his criticism of science. He alleges against science its absorption in abstractions. His consistent theme is that the important facts of nature elude the scientific method. It is important therefore to ask what Wordsworth found in nature that failed to receive expression in science. I ask this question in the interest of science itself; for one main position in these lectures is a protest against the idea that the abstractions of science are irreformable and unalterable. Now it is emphatically not the case that Wordsworth hands over inorganic matter to the mercy of science, and concentrates on the faith that in the living organism there is some element that science cannot analyse. Of course he recognizes, what no one doubts, that in some sense living things are different from lifeless things. But that is not his main point. It is the brooding presence of the hills which haunts him. His theme is nature *in solido*, that is to say, he dwells on that mysterious presence of surrounding things, which imposes itself on any separate element that we set up as an individual for its own sake. He always grasps the whole of nature as involved in the tonality of the particular instance. That is why he laughs with the daffodils, and finds in the primrose 'thoughts too deep for tears'.

Wordsworth's greatest poem is, by far, the first book of *The Prelude*. It is pervaded by this sense of the haunting presences of nature. A series of magnificent passages, too long for quotation, express this idea. Of course, Wordsworth is a poet writing a poem, and is not concerned with dry philosophical statements. But it would hardly be possible to express more clearly a feeling for nature, as exhibiting entwined prehensive unities, each suffused with modal presences of others:

> Ye Presences of Nature in the sky
> And on the earth! Ye Visions of the hills!
> And Souls of lonely places! can I think
> A vulgar hope was yours when ye employed
> Such ministry, when ye through many a year
> Haunting me thus among my boyish sports,
> On caves and trees, upon the woods and hills,
> Impressed upon all forms the characters
> Of danger or desire; and thus did make

The surface of the universal earth
With triumph and delight, with hope and fear,
Work like a sea? . . .

In thus citing Wordsworth, the point which I wish to make is that we forget how strained and paradoxical is the view of nature which modern science imposes on our thoughts. Wordsworth, to the height of genius, expresses the concrete facts of our apprehension, facts which are distorted in the scientific analysis. Is it not possible that the standardized concepts of science are only valid within narrow limitations, perhaps too narrow for science itself?

Shelley's attitude to science was at the opposite pole to that of Wordsworth. He loved it, and is never tired of expressing in poetry the thoughts which it suggests. It symbolizes to him joy, and peace, and illumination. What the hills were to the youth of Wordsworth, a chemical laboratory was to Shelley. It is unfortunate that Shelley's literary critics have, in this respect, so little of Shelley in their own mentality. They tend to treat as a casual oddity of Shelley's nature what was, in fact, part of the main structure of his mind, permeating his poetry through and through. If Shelley had been born a hundred years later, the twentieth century would have seen a Newton among chemists.

For the sake of estimating the value of Shelley's evidence it is important to realize this absorption of his mind in scientific ideas. It can be illustrated by lyric after lyric. I will choose one poem only, the fourth act of his *Prometheus Unbound*. The earth and the Moon converse together in the language of accurate science. Physical experiments guide his imagery. For example, the Earth's exclamation,

The vaporous exultation not to be confined!

is the poetic transcript of 'the expansive force of gases', as it is termed in books on science. Again, take the Earth's stanza,

I spin beneath my pyramid of night,
Which points into the heavens, – dreaming delight,
Murmuring victorious joy in my enchanted sleep;
As a youth lulled in love-dreams faintly sighing,
Under the shadow of his beauty lying,
Which round his rest a watch of light and warmth doth
  keep.

This stanza could only have been written by someone with a definite geometrical diagram before his inward eye – a diagram which it has often been my business to demonstrate to mathematical classes. As evidence, note especially the last line which gives poetical imagery to the light surrounding night's pyramid. This idea could not occur to anyone without the diagram. But the whole poem and other poems are permeated with touches of this kind.

Now the poet, so sympathetic with science, so absorbed in its ideas, can simply make nothing of the doctrine of secondary qualities which is fundamental to its concepts. For Shelley nature retains its beauty and its colour. Shelley's nature is in its essence a nature of organisms, functioning with the full content of our perceptual experience. We are so used to ignoring the implication of orthodox scientific doctrine, that it is difficult to make evident the criticism upon it which is thereby implied. If anybody could have treated it seriously, Shelley would have done so.

Furthermore Shelley is entirely at one with Wordsworth as to the interfusing of the presence in nature. Here is the opening stanza of his poem entitled *Mont Blanc*:

> The everlasting universe of Things
> Flows through the Mind, and rolls its rapid waves,
> Now dark – now glittering – now reflecting gloom –
> Now lending splendour, where from secret springs
> The source of human thought its tribute brings
> Of waters, – with a sound but half its own,

> Such as a feeble brook will oft assume
> In the wild woods, among the Mountains lone,
> Where waterfalls around it leap for ever,
> Where woods and winds contend, and a vast river
> Over its rocks ceaselessly bursts and raves.

Shelley has written these lines with explicit reference to some form of idealism, Kantian or Berkeleyan or Platonic. But however you construe him, he is here an emphatic witness to a prehensive unification as constituting the very being of nature.

Berkeley, Wordsworth, Shelley are representative of the intuitive refusal seriously to accept the abstract materialism of science.

There is an interesting difference in the treatment of nature by Wordsworth and by Shelley, which brings forward the exact questions we have got to think about. Shelley thinks of nature as changing, dissolving, transforming as it were at a fairy's touch. The leaves fly before the west wind

> Like ghosts from an enchanter fleeing.

In his poem *The Cloud* it is the transformations of water which excite his imagination. The subject of the poem is the endless, eternal, elusive change of things:

> I change but I cannot die.

This is one aspect of nature, its elusive change: a change not merely to be expressed by locomotion, but a change of inward character. This is where Shelley places his emphasis, on the change of what cannot die.

Wordsworth was born among hills; hills mostly barren of trees, and thus showing the minimum of change with the seasons. He was haunted by the enormous permanences of nature. For him change is an incident which shoots across a background of endurance,

> Breaking the silence of the seas
> Among the farthest Hebrides.

Every scheme for the analysis of nature has to face these two facts, change and endurance. There is yet a third fact to be placed by it, eternality, I will call it. The mountain endures. But when after ages it has been worn away, it has gone. If a replica arises, it is yet a new mountain. A colour is eternal. It haunts time like a spirit. It comes and it goes. But where it comes, it is the same colour. It neither survives nor does it live. It appears when it is wanted. The mountain has to time and space a different relation from that which colour has. In the previous lecture, I was chiefly considering the relation to space-time of things which, in my sense of the term, are eternal. It was necessary to do so before we can pass to the consideration of the things which endure.

Also we must recollect the basis of our procedure. I hold that philosophy is the basis of abstractions. Its function is the double one, first of harmonizing them by assigning to them their right relative status as abstractions, and secondly of completing them by direct comparison with more concrete intuitions of the universe, and thereby promoting the formation of more complete schemes of thought. It is in respect to this comparison that the testimony of great poets is of such importance. Their survival is evidence that they express deep intuitions of mankind penetrating into what is universal in concrete fact. Philosophy is not one among the sciences with its own little scheme of abstractions which it works away at perfecting and improving. It is the survey of sciences, with the special objects of their harmony, and of their completion. It brings to this task, not only the evidence of the separate sciences, but also its own appeal to concrete experience. It confronts the sciences with concrete fact.

The literature of the nineteenth century, especially its English poetic literature, is a witness to the discord between the aesthetic intuitions of mankind and the mechanism of science. Shelley brings vividly before us the elusiveness of the eternal objects of sense as they haunt the change which infects underlying organisms. Wordsworth is the poet of nature as being the field of enduring permanences carrying within themselves a message of tremendous significance. The eternal objects are also there for him,

> The light that never was, on sea or land.

Both Shelley and Wordsworth emphatically bear witness that nature cannot be divorced from its aesthetic values; and that these values arise from the cumulation, in some sense, of the brooding presence of the whole on to its various parts. Thus we gain from the poets the doctrine that a philosophy of nature must concern itself at least with these five notions: change, value, eternal objects, endurance, organism, interfusion.

We see that the literary romantic movement at the beginning of the nineteenth century, just as much as Berkeley's philosophical idealistic movement a hundred years earlier, refused to be confined within the materialistic concepts of the orthodox scientific theory. We know also that when in these lectures we come to the twentieth century, we shall find a movement in science itself to reorganize its concepts, driven thereto by its own intrinsic development.

It is, however, impossible to proceed until we have settled whether this refashioning of ideas is to be carried out on an objectivist basis or on a subjectivist basis. By a subjectivist basis I mean the belief that the nature of our immediate experience is the outcome of the perceptive peculiarities of the subject enjoying the experience. In other words, I mean that for this theory what is perceived is not a partial vision of a complex of things generally independent of that act of cognition; but that it merely is the expression of the individual peculiarities of the cognitive act. Accordingly what is common to the multiplicity of cognitive acts is the ratiocination connected with them. Thus, though there is a common world of thought associated with our sense-perceptions, there is no common world to think about. What we do think about is a common conceptual world applying indifferently to our individual experiences which are strictly personal to ourselves. Such a conceptual world will ultimately find its complete expression in the equations of applied mathematics. This is the extreme subjectivist position. There is of course the half-way house of those who believe that our perceptual experience does tell us of a common objective world; but that the things perceived are merely the outcome for us of this world, and are not in themselves elements in the common world itself.

Also there is the objectivist position. This creed is that the actual elements perceived by our senses are in themselves the elements of a common world; and that this world is a complex of things, including indeed our acts of cognition. There is of course the half-way house of those who beview the things experienced are to be distinguished from our knowledge of them. So far as there is dependence, the things pave the way for the cognition, rather than vice versa. But the point is that the actual things experienced enter into a common world which transcends knowledge, though it includes knowledge. The intermediate subjectivists would hold that the things experienced only indirectly enter into the common world by reason of their dependence on the subject who is cognizing. The objectivist holds that the things experienced and the cognizant subject enter into the common world on equal terms. In these lectures I am giving the outline of what I consider to be the essentials of an objectivist philosophy adapted to the requirement of science and to the concrete experience of mankind. Apart from the detailed criticism of the difficulties raised by subjectivism in any form, my broad reasons for distrusting it are three in number. One reason arises from the direct interrogation of our perceptive inexperience. It appears from this interrogation that we are *within* a world of colours, sounds, and other sense-objects, related in space and time to enduring objects such as stones, trees, and human bodies. We seem to be ourselves elements of this world in the same sense as are the other things which we perceive. But the subjectivist, even the moderate intermediate subjectivist, makes this world, as thus described, depend on us, in a way which directly traverses our naïve experience. I hold that the ultimate appeal is to naïve experience and that is why I lay such stress on the evidence of poetry. My point is, that in our sense-experience we know away from and beyond our own personality; whereas the subjectivist holds that in such experience we merely know about our own personality. Even the intermediate subjectivist places our personality between the world we know of and the common world which he admits. The world we know of is for him the internal strain of our personality under the stress of the common world which lies behind.

My second reason for distrusting subjectivism is based on the particular content of experience. Our historical knowledge tells us of ages in the past when, so far as we can see, no living being existed on earth. Again it also tells us of countless star-systems, whose detailed history remains beyond our ken. Consider even the moon and the earth. What is going on within the interior of the earth, and on the far side of the moon! Our perceptions lead us to infer that there is something happening in the stars, something happening within the earth, and something happening on the far side of the moon. Also they tell us that in remote ages there were things happening. But all these things which it appears certainly happened, are either unknown in detail, or else are reconstructed by inferential evidence. In the face of this content of our personal experience, it is difficult to believe that the experienced world is an attribute of our own personality. My third reason is based upon the instinct for action. Just as sense-perception seems to give knowledge of what lies beyond individuality, so action seems to issue in an instinct for self-transcendence. The activity passes beyond self into the known transcendent world. It is here that final ends are of importance. For it is not activity urged from behind, which passes out into the veiled world of the intermediate subjectivist. It is activity directed to determinate ends in the known world; and yet it is activity transcending self and it is activity within the known world. It follows therefore that the world, as known, transcends the subject which is cognizant of it.

The subjectivist position has been popular among those who have been engaged in giving a philosophical interpretation to the recent theories of relativity in physical science. The dependence of the world of sense on the individual percipient seems an easy mode of expressing the meanings involved. Of course, with the exception of those who are content with themselves as forming the entire universe, solitary amid nothing, everyone wants to struggle back to some sort of objectivist position. I do not understand how a common world of thought can be established in the absence of a common world of sense. I will not argue this point in detail; but in the absence of a transcendence of thought, or a transcendence of the world of sense, it is difficult to see how the subjectivist is to divest himself of his solitariness. Nor does the intermediate subjectivist appear to get any help from his unknown world in the background.

The distinction between realism and idealism does not coincide with that between objectivism and subjectivism. Both realists and idealists can start from an objective standpoint. They may both agree that the world disclosed in sense-perception is a common world, transcending the individual recipient. But the objective idealist, when he comes to analyse what the reality of this world involves, finds that cognitive mentality is in some way inextricably concerned in every detail. This position the realist denies. Accordingly, these two classes of objectivists do not part company till they have arrived at the ultimate problem of metaphysics. There is a great deal which they share in common. This is why, in my last lecture, I said that I adopted a position of provisional realism.

In the past, the objectivist position has been distorted by the supposed necessity of accepting the classical scientific materialism, with its doctrine of simple location. This has necessitated the doctrine of secondary and primary qualities. Thus the secondary qualities, such as the sense-objects, are dealt with on subjectivist principles. This is a half-hearted position which falls an easy prey to subjectivist criticism.

If we are to include the secondary qualities in the common world, a very drastic reorganization of our fundamental concept is necessary. It is an evident fact of experience that our apprehensions of the external world depend absolutely on the occurrences within the human body. By playing appropriate tricks on the body a man can be got to perceive, or not to perceive, almost anything. Some people express themselves as though bodies, brains, and nerves were the only real things in an entirely imaginary world. In other words, they treat bodies on objectivist principles, and the rest of the world on subjectivist principles. This will not do; especially when we remember that it is the experimenter's perception of another person's body which is in question as evidence.

But we have to admit that the body is the organism whose states regulate our cognizance of the world. The unity of the perceptual field therefore must be a unity of bodily experience. In being aware of the bodily experience, we must thereby be aware of aspects of the whole spatio-temporal world as mirrored within the bodily life. This is the solution of the problem which I gave in my last lecture. I will not repeat myself now, except to remind you that my theory involves the entire abandonment of the notion that simple location is the primary way in which things are involved in space-time. In a certain sense, everything is everywhere at all times. For every location involves an aspect of itself in every other location. Thus every spatio-temporal standpoint mirrors the world.

If you try to imagine this doctrine in terms of our conventional views of space and time, which presuppose simple location, it is a great paradox. But if you think of it in terms of our naïve experience, it is a mere transcript of the obvious facts. You are in a certain place perceiving things. Your perception takes place where you are, and is entirely dependent on how your body is functioning. But this functioning of the body in one place, exhibits for your cognizance an aspect of the distant environment, fading away into the general knowledge that there are things beyond. If this cognizance conveys knowledge of a transcendent world, it must be because the event which is the bodily life unifies in itself aspects of the universe.

This is a doctrine extremely consonant with the vivid expression of personal experience which we find in the nature-poetry of imaginative writers such as Wordsworth or Shelley. The brooding, immediate presences of things are an obsession to Wordsworth. What the theory does do is to edge cognitive mentality away from being the necessary substratum of the unity of experience. That unity is now placed in the unity of an event. Accompanying this unity, there may or there may not be cognition.

At this point we come back to the great question which was posed before us by our examination of the evidence afforded by the poetic insight of Wordsworth and Shelley. This single question has expanded into a group of questions. What are enduring things, as distinguished from the eternal objects, such as colour and shape? How are they possible? What is their status and meaning in the universe? It comes to this: What is the status of the enduring stability of the order of nature? There is the summary answer, which refers nature to some greater reality standing behind it. This reality occurs in the history of thought under many names, the absolute, Brahma, the order of Heaven, God.

The delineation of final metaphysical truth is no part of this lecture. My point is that any summary conclusion jumping from our conviction of the existence of such an order of nature to the easy assumption that there is an ultimate reality which, in some unexplained way, is to be appealed to for the removal of perplexity, constitutes the great refusal of rationality to assert its rights. We have to search whether nature does not in its very being show itself as self-explanatory. By this I mean, that the sheer statement, of what things are, may contain elements explanatory of why things are. Such elements may be expected to refer to depths beyond anything which we can grasp with a clear apprehension. In a sense, all explanation must end in an ultimate arbitrariness. My demand is, that the ultimate arbitrariness of matter of fact from which our formulation starts should disclose the same general principles of reality, which we dimly discern as stretching away into regions beyond our explicit powers of discernment. Nature exhibits itself as exemplifying a philosophy of the evolution of organisms subject to determinate conditions. Examples of such conditions are the dimensions of space, the laws of nature, the determinate enduring entities, such as atoms and electrons, which exemplify these laws. But the very nature of these entities, the very nature of their spatiality and temporality, should exhibit the arbitrariness of these conditions as the outcome of a wider evolution beyond nature itself, and within which nature is but a limited mode.

One all-pervasive fact, inherent in the very character of what is real, is the transition of things, the passage one to another. This passage is not a mere linear procession of discrete entities. However we fix a determinate entity, there is always a narrower determination of something which is presupposed in our first choice. Also there is always a wider determination into which our first choice fades by transition beyond itself. The general aspect of nature is that of evolutionary expansiveness. These unities, which I call events, are the emergence into actuality of something. How are we to characterize the something which thus emerges? The name 'event' given to such a unity, draws attention to the inherent transitoriness, combined with the actual unity. But this abstract word cannot be sufficient to characterize what the fact of the reality of an event is in itself. A moment's thought shows us that no one idea can in itself be sufficient. For every idea which finds its significance in each event must represent something which contributes to what realization is in itself. Thus no one word can be adequate. But conversely, nothing must be left out. Remembering the poetic rendering of our concrete experience, we see at once that the element of value, of being valuable, of having value, of being an end in itself, of being something which is for its own sake, must not be omitted in any account of an event as the most concrete actual something. 'Value' is the word I use for the intrinsic reality of an event. Value is an element which permeates through and through the poetic view of nature. We have only to transfer to the very texture of realization in itself that value which we recognize so readily in terms of human life. This is the secret of Wordsworth's worship of nature. Realization therefore is in itself the attainment of value. But there is no such thing as mere value. Value is the outcome of limitation. The definite finite entity is the selected mode which is the shaping of attainment; apart from such shaping into individual matter of fact there is no attainment. The mere fusion of all that there is would be the nonentity of indefiniteness. The salvation of reality is its obstinate, irreducible, matter-of-fact entities, which are limited to be no other than themselves. Neither science, nor art, nor creative action can tear itself away from obstinate, irreducible, limited facts. The endurance of things has its significance in the self-retention of that which imposes itself as a definite attainment for its own sake. That which endures is limited, obstructive, intolerant, infecting its environment with its own aspects. But it is not self-sufficient. The aspects of all things enter into its very nature. It is only itself as drawing together into its own limitation the larger whole in which it finds itself. Conversely it is only itself by lending its aspects to this same environment in which it finds itself. The problem of evolution is the development of enduring harmonies of enduring shapes of value, which merge into higher attainments of things beyond themselves. Aesthetic attainment is interwoven in the texture of realization. The endurance of an entity represents the attainment of a limited aesthetic success, though if we look beyond it to its external effects, it may represent an aesthetic failure. Even within itself, it may represent the conflict between a lower success and a higher failure. The conflict is the presage of disruption.

The further discussion of the nature of enduring objects and of the conditions they require will be relevant to the consideration of the doctrine of evolution which dominated the latter half of the nineteenth century. The point which in this lecture I have endeavoured to make clear is that the nature poetry of the romantic revival was a protest on behalf of the organic view of nature, and also a protest against the exclusion of value from the essence of matter of fact. In this aspect of it, the romantic movement may be conceived as a revival of Berkeley's protest which had been launched a hundred years earlier. The romantic reaction was a protest on behalf of value.

CHAPTER VI

THE NINETEENTH CENTURY

My previous lecture was occupied with the comparison of the nature-poetry of the romantic movement in England with the materialistic scientific philosophy inherited from the eighteenth century. It noted the entire disagreement of the two movements of thought. The lecture also continued the endeavour to outline an objectivist philosophy, capable of bridging the gap between science and that fundamental intuition of mankind which finds its expression in poetry and its practical exemplification in the presuppositions of daily life. As the nineteenth century passed on, the romantic movement died down. It did not die away, but it lost its clear unity of tidal stream, and dispersed itself into many estuaries as it coalesced with other human interests. The faith of the century was derived from three sources: one source was the romantic movement, showing itself in religious revival, in art, and in political aspiration: another source was the gathering advance of science which opened avenues of thought: the third source was the advance in technology which completely changed the conditions of human life.

Each of these springs of faith had its origin in the previous period. The French Revolution itself was the first child of romanticism in the form in which it tinged Rousseau. James Watt obtained his patent for his steam-engine in 1769. The scientific advance was the glory of France and of French influence, throughout the same century.

Also even during this earlier period, the streams interacted, coalesced, and antagonized each other. But it was not until the nineteenth century that the threefold movement came to that full development and peculiar balance characteristic of the sixty years following the battle of Waterloo.

What is peculiar and new to the century, differentiating it from all its predecessors, is its technology. It was not merely the introduction of some great isolated inventions. It is impossible not to feel that something more than that was involved. For example, writing was a greater invention than the steam-engine. But in tracing the continuous history of the growth of writing we find an immense difference from that of the steam-engine. We must, of course, put aside minor and sporadic anticipations of both; and confine attention to the periods of their effective elaboration. For scale of time is so absolutely disparate. For the steam-engine, we may give about a hundred years; for writing, the time period is of the order of a thousand years. Further, when writing was finally popularized, the world was not then expecting the next step in technology. The process of change was slow, unconscious, and unexpected.

In the nineteenth century, the process became quick, conscious, and expected. The earlier half of the century was the period in which this new attitude to change was first established and enjoyed. It was a peculiar period of hope, in the sense in which, sixty or seventy years later, we can now detect a note of disillusionment, or at least of anxiety.

The greatest invention of the nineteenth century was the invention of the method of invention. A new method entered into life. In order to understand our epoch, we can neglect all the details of change, such as railways, telegraphs, radios, spinning machines, synthetic dyes. We must concentrate on the method in itself; that is the real novelty, which has broken up the foundations of the old civilization. The prophecy of Francis Bacon has now been fulfilled; and man, who at times dreamt of himself as a little lower than the angels, has submitted to become the servant and the minister of nature. It still remains to be seen whether the same actor can play both parts.

The whole change has arisen from the new scientific information. Science, conceived not so much in its principles as in its results, is an obvious store-house of ideas for utilization. But, if we are to understand what happened during the century, the analogy of a mine is better than that of a store-house. Also, it is a great mistake to think that the bare scientific idea is the required invention, so that it has only to be picked up and used. An intense period of imaginative design lies between. One element in the new method is just the discovery of how to set about bridging the gap between the scientific ideas, and the ultimate product. It is a process of disciplined attack upon one difficulty after another.

The possibilities of modern technology were first in practice realized in England, by the energy of a prosperous middle class. Accordingly, the industrial revolution started there. But the Germans explicitly realized the methods by which the deeper veins in the mine of science could be reached. They abolished haphazard methods of scholarship. In their technological schools and universities progress did not have to wait for the occasional genius, or the occasional lucky thought. Their feats of scholarship during the nineteenth century were the admiration of the world. This discipline of knowledge applies technology to pure science, and beyond science to general scholarship. It represents the change from amateurs to professionals.

There have always been people who devoted their lives to definite regions of thought. In particular, lawyers and the clergy of the Christian churches form obvious examples of such specialism. But the full self-conscious realization of the power of professionalism in knowledge in all its departments, and of the way to produce the professionals, and of the importance of knowledge to the advance of technology, and of the methods by which abstract knowledge can be connected with technology, and of the boundless possibilities of technological advance – the realization of all these things was first completely attained in the nineteenth century; and among the various countries, chiefly in Germany.

In the past human life was lived in a bullock cart; in the future it will be lived in an aeroplane; and the change of speed amounts to a difference in quality.

The transformation of the field of knowledge, which has been thus effected, has not been wholly a gain. At least, there are dangers implicit in it, although the increase of efficiency is undeniable. The discussion of various effects on social life arising from the new situation is reserved for my last lecture. For the present it is sufficient to note that this novel situation of disciplined progress is the setting within which the thought of the century developed.

In the period considered four great novel ideas were introduced into theoretical science. Of course, it is possible to show good cause for increasing my list far beyond the number four. But I am keeping to ideas which, if taken in their broadest signification, are vital to modern attempts at reconstructing the foundations of physical science.

Two of these ideas are antithetical, and I will consider them together. We are not concerned with details, but with ultimate influences on thought. One of the ideas is that of a field of physical activity pervading all space, even where there is an apparent vacuum. This notion had occurred to many people, under many forms. We remember the medieval axiom, nature abhors a vacuum. Also, Descartes' vortices at one time, in the seventeenth century, seemed as if established among scientific assumptions. Newton believed that gravitation was caused by something happening in a medium. But, on the whole, in the eighteenth century nothing was made of any of these ideas. The passage of light was explained in Newton's fashion by the flight of minute corpuscles, which of course left room for a vacuum. Mathematical physicists were far too busy deducing the consequences of the theory of gravitation to bother much about the causes; nor did they know where to look, if they had troubled themselves over the question. There were speculations, but their importance was not great. Accordingly, when the nineteenth century opened, the notion of physical occurrences pervading all space held no effective place in science. It was revived from two sources. The undulatory theory of light triumphed, thanks to Thomas Young and Fresnel. This demands that there shall be something throughout space which can undulate. Accordingly, the ether was produced, as a sort of all pervading subtle material. Again the theory of electromagnetism finally, in Clerk Maxwell's hands, assumed a shape in which it demanded that there should be electromagnetic occurrences throughout all space. Maxwell's complete theory was not shaped until the eighteen-seventies. But it had been prepared for by many great men, Ampère, Oersted, Faraday. In accordance with the current materialistic outlook, these electromagnetic occurrences also required a material in which to happen. So again the ether was requisitioned. Then Maxwell, as the immediate first-fruits of his theory, demonstrated that the waves of light were merely waves of his electromagnetic occurrences. Accordingly, the theory of electromagnetism swallowed up the theory of light. It was a great simplification, and no one doubts its truth. But it had one unfortunate effect so far as materialism was concerned. For, whereas quite a simple sort of elastic ether sufficed for light when taken by itself, the electromagnetic ether has to be endowed with just those properties necessary for the production of the electromagnetic occurrences. In fact, it becomes a mere name for the material which is postulated to underlie these occurrences. If you do not happen to hold the metaphysical theory which makes you postulate such an ether, you can discard it. For it has no independent vitality.

Thus in the seventies of the last century, some main physical sciences were established on a basis which presupposed the idea of continuity. On the other hand, the idea of atomicity had been introduced by John Dalton, to complete Lavoisier's work on the foundation of chemistry. This is the second great notion. Ordinary matter was conceived as atomic: electromagnetic effects were conceived as arising from a continuous field.

There was no contradiction. In the first place, the notions are antithetical; but, apart from special embodiments, are not logically contradictory. Secondly, they were applied to different regions of science, one to chemistry, and the other to electromagnetism. And, as yet, there were but faint signs of coalescence between the two.

The notion of matter as atomic has a long history. Democritus and Lucretius will at once occur to your minds. In speaking of these ideas as novel, I merely mean relatively novel, having regard to the settlement of ideas which formed the efficient basis of science throughout the eighteenth century. In considering the history of thought, it is necessary to distinguish the real stream, determining a period, from ineffectual thoughts casually entertained. In the eighteenth century every well-educated man read Lucretius, and entertained ideas about atoms. But John Dalton made them efficient in the stream of science; and in this function of efficiency atomicity was a new idea.

The influence of atomicity was not limited to chemistry. The living cell is to biology what the electron and the proton are to physics. Apart from cells and from aggregates of cells there are no biological phenomena. The cell theory was introduced into biology contemporaneously with, and independently of, Dalton's atomic theory. The two theories are independent exemplifications of the same idea of 'atomism'. The biological cell theory was a gradual growth, and a mere list of dates and names illustrates the fact that the biological sciences, as effective schemes of thought, are barely one hundred years old. Bichât in 1801 elaborated a tissue theory: Johannes Müller in 1835 described 'cells' and demonstrated facts concerning their nature and relations: Schleiden in 1838 and Schwann in 1839 finally established their fundamental character. Thus by 1840 both biology and chemistry were established on an atomic basis.

The final triumph of atomism had to wait for the arrival of electrons at the end of the century. The importance of the imaginative background is illustrated by the fact that nearly half a century after Dalton had done his work, another chemist, Louis Pasteur, carried over these same ideas of atomicity still further into the region of biology. The cell theory and Pasteur's work were in some respects more revolutionary than that of Dalton. For they introduced the notion of organism into the world of minute beings. There had been a tendency to treat the atom as an ultimate entity, capable only of external relations. This attitude of mind was breaking down under the influence of Mendeleëf's periodic law. But Pasteur showed the decisive importance of the idea of organism at the stage of infinitesimal magnitude. The astronomers had shown us how big is the universe. The chemists and biologists teach us how small it is. There is in modern scientific practice a famous standard of length. It is rather small : to obtain it, you must divide a centimetre into one hundred million parts, and take one of them. Pasteur's organisms are a good deal bigger than this length. In connection with atoms, we now know that there are organisms for which such distances are uncomfortably great.

The remaining pair of new ideas to be ascribed to this epoch are both of them connected with the notion of transition or change. They are the doctrine of the conservation of energy, and the doctrine of evolution.

The doctrine of energy has to do with the notion of quantitative permanence underlying change. The doctrine of evolution has to do with the emergence of novel organisms as the outcome of change. The theory of energy lies in the province of physics. The theory of evolution lies mainly in the province of biology, although it had previously been touched upon by Kant and Laplace in connection with the formation of suns and planets.

The convergent effect of the new power for scientific advance, which resulted from these four ideas, transformed the middle period of the century into an orgy of scientific triumph. Clear-sighted men, of the sort who are so clearly wrong, now proclaimed that the secrets of the physical universe were finally disclosed. If only you ignored everything which refused to come into line, your powers of explanation were unlimited. On the other side, muddle-headed men muddled themselves into the most indefensible positions. Learned dogmatism, conjoined with ignorance of the crucial facts, suffered a heavy defeat from the scientific advocates of new ways. Thus to the excitement derived from technological revolution, there was now added the excitement arising from the vistas disclosed by scientific theory. Both the material and the spiritual bases of social life were in process of transformation. When the century entered upon its last quarter, its three sources of inspiration, the romantic, the technological, and the scientific had done their work.

Then, almost suddenly, a pause occurred; and in its last twenty years the century closed with one of the dullest stages of thought since the time of the First Crusade. It was an echo of the eighteenth century, lacking Voltaire and the reckless grace of the French aristocrats. The period was efficient, dull, and half-hearted. It celebrated the triumph of the professional man.

But looking backwards upon this time of pause, we can now discern signs of change. In the first place, the modern conditions of systematic research prevent absolute stagnation. In every branch of science, there was effective progress, indeed rapid progress, although it was confined somewhat strictly within the accepted ideas of each branch. It was an age of successful scientific orthodoxy, undisturbed by much thought beyond the conventions.

In the second place, we can now see that the adequacy of scientific materialism as a scheme of thought for the use of science was endangered. The conservation of energy provided a new type of quantitative permanence. It is true that energy could be construed as something subsidiary to matter. But, anyhow, the notion of mass was losing its unique pre-eminence as being the one final permanent quantity. Later on, we find the relations of mass and energy inverted; so that mass now becomes the name for a quantity of energy considered in relation to some of its dynamical effects. This train of thought leads to the notion of energy being fundamental, thus displacing matter from that position. But energy is merely the name for the quantitative aspect of a structure of happenings; in short, it depends on the notion of the functioning of an organism. The question is, can we define an organism without recurrence to the concept of matter in simple location? We must, later on, consider this point in more detail.

The same relegation of matter to the background occurs in connection with the electromagnetic fields. The modern theory presupposes happenings in that field which are divorced from immediate dependence upon matter. It is usual to provide an ether as a substratum. But the ether does not really enter into the theory. Thus again the notion of material loses its fundamental position. Also, the atom is transforming itself into an organism; and finally the evolution theory is nothing else than the analysis of the conditions for the formation and survival of various types of organisms. In truth, one most significant fact of this later period is the advance in biological sciences. These sciences are essentially sciences concerning organisms. During the epoch in question, and indeed also at the present moment, the prestige of the more scientific form belongs to the physical sciences. Accordingly, biology apes the manners of physics. It is orthodox to hold that there is nothing in biology but what is physical mechanism under somewhat complex circumstances.

One difficulty in this position is the present confusion as to the foundational concepts of physical science. This same difficulty also attaches to the opposed doctrine of vitalism. For, in this later theory, the fact of mechanism is accepted – I mean, mechanism based upon materialism – and an additional vital control is introduced to explain the actions of living bodies. It cannot be too clearly understood that the various physical laws which appear to apply to the behaviour of atoms are not mutually consistent as at present formulated. The appeal to mechanism on behalf of biology was in its origin an appeal to the well-attested self-consistent physical concepts as expressing the basis of all natural phenomena. But at present there is no such system of concepts.

Science is taking on a new aspect which is neither purely physical, nor purely biological. It is becoming the study of organisms. Biology is the study of the larger organisms; whereas physics is the study of the smaller organisms. There is another difference between the two divisions of science. The organisms of biology include as ingredients the smaller organisms of physics; but there is at present no evidence that the smaller of the physical organisms can be analysed into component organisms. It may be so. But anyhow we are faced with the question as to whether there are not primary organisms which are incapable of further analysis. It seems very unlikely that there should be any infinite regress in nature. Accordingly, a theory of science which discards materialism must answer the question as to the character of these primary entities. There can be only one answer on this basis. We must start with the event as the ultimate unit of natural occurrence. An event has to do with all that there is, and in particular with all other events. This interfusion of events is effected by the aspects of those eternal objects, such as colours, sounds, scents, geometrical characters, which are required for nature and are not emergent from it. Such an eternal object will be an ingredient of one event under the guise, or aspect, of qualifying another event. There is a reciprocity of aspects, and there are patterns of aspects. Each event corresponds to two such patterns; namely, the pattern of aspects of other events which it grasps into its own unity, and the pattern of its aspects which other events severally grasp into their unities. Accordingly, a non-materialistic philosophy of nature will identify a primary organism as being the emergence of some particular pattern as grasped in the unity of a real event. Such a pattern will also include the aspects of the event in question as grasped in other events, whereby those other events receive a modification, or partial determination. There is thus an intrinsic and an extrinsic reality of an event, namely, the event as in its own prehension, and the event as in the prehension of other events. The concept of an organism includes, therefore, the concept of the interaction of organisms. The ordinary scientific ideas of transmission and continuity are, relatively speaking, details concerning the empirically observed characters of these patterns throughout space and time. The position here maintained is that the relationships of an event are internal, so far as concerns the event itself; that is to say, that they are constitutive of what the event is in itself.

Also in the previous lecture, we arrived at the notion that an actual event is an achievement for its own sake, a grasping of diverse entities into a value by reason of their real togetherness in that pattern, to the exclusion of other entities. It is not the mere logical togetherness of merely diverse things. For in that case, to modify Bacon's words, 'all eternal objects would be alike one to another.' This reality means that each intrinsic essence, that is to say, what each eternal object is in itself, becomes relevant to the one limited value emergent in the guise of the event. But values differ in importance. Thus though each event is necessary for the community of events, the weight of its contribution is determined by something intrinsic in itself. We have now to discuss what that property is. Empirical observation shows that it is the property which we may call indifferently retention, endurance of reiteration. This property amounts to the recovery, on behalf of value amid the transitoriness of reality, of the self-identity which is also enjoyed by the primary eternal objects. The reiteration of a particular shape (or formation) of value within an event occurs when the event as a whole repeats some shape which is also exhibited by each one of a succession of its parts. Thus, however you analyse the event according to the flux of its parts through time, there is the same thing-for-its-own-sake standing before you. Thus the event, in its own intrinsic reality, mirrors in itself, as derived from its own parts, aspects of the same patterned value as it realizes in its complete self. It thus realizes itself under the guise of an enduring individual entity, with a life-history contained within itself. Furthermore, the extrinsic reality of such an event, as mirrored in other events, takes this same form of an enduring individuality; only in this case, the individuality is implanted as a reiteration of aspects of itself in the alien events composing the environment.

The total temporal duration of such an event bearing an enduring pattern, constitutes its specious present. Within this specious present the event realizes itself as a totality, and also in so doing realizes itself as grouping together a number of aspects of its own temporal parts. One and the same pattern is realized in the total event, and is exhibited by each of these various parts through an aspect of each part grasped into the togetherness of the total event. Also, the earlier life-history of the same pattern is exhibited by its aspects in this total event. There is, thus, in this event a memory of the antecedent life-history of its own dominant pattern, as having formed an element of value in its own antecedent environment. This concrete prehension, from within, of the life-history of an enduring fact is analysable into two abstractions, of which one is the enduring entity which has emerged as a real matter of fact to be taken account of by other things, and the other is the individualized embodiment of the underlying energy of realization.

The consideration of the general flux of events leads to this analysis into an underlying eternal energy in whose nature there stands an envisagement of the realm of all eternal objects. Such an envisagement is the ground of the individualized thoughts which emerge as thought-aspects grasped within the life-history of the subtler and more complex enduring patterns. Also in the nature of the eternal activity there must stand an envisagement of all values to be obtained by a real togetherness of eternal objects, as envisaged in ideal situations. Such ideal situations, apart from any reality, are devoid of intrinsic value, but are valuable as elements in purpose. The individualized prehension into individual events of aspects of these ideal situations takes the form of individualized thoughts, and as such has intrinsic value. Thus value arises because there is now a real togetherness of the idea aspects, as in thought, with the actual aspects, as in process of occurrence. Accordingly no value is to be ascribed to the underlying activity as divorced from the matter-of-fact events of the real world.

Finally, to sum up this train of thought, the underlying activity, as conceived apart from the fact of realization, has three types of envisagement. These are: first, the envisagement of eternal objects; secondly, the envisagement of possibilities of value in respect to the synthesis of eternal objects; and lastly, the envisagement of the actual matter of fact which must enter into the total situation which is achievable by the addition of the future. But in abstraction from actuality, the eternal activity is divorced from value. For the actuality is the value. The individual perception arising from enduring objects will vary in its individual depth and width according to the way in which the pattern dominates its own route. It may represent the faintest ripple differentiating the general substrate energy; or, in the other extreme, it may rise to conscious thought, which includes poising before self-conscious judgement the abstract possibilities of value inherent in various situations of ideal togetherness. The intermediate cases will group round the individual perception as envisaging (without self-consciousness) that one immediate possibility of attainment which represents the closest analogy to its own immediate past, having regard to the actual aspects which are there for prehension. The laws of physics represent the harmonized adjustment of development which results from this unique principle of determination. Thus dynamics is dominated by a principle of least action, whose detailed character has to be learnt from observation.

The atomic material entities which are considered in physical science are merely these individual enduring entities, conceived in abstraction from everything except what concerns their mutual interplay in determining each other's historical routes of life-history. Such entities are partially formed by the inheritance of aspects from their own past. But they are also partially formed by the aspects of other events forming their environments. The laws of physics are the laws declaring how the entities mutually react among themselves. For physics these laws are arbitrary, because that science has abstracted from what the entities are in themselves. We have seen that this fact of what the entities are in themselves is liable to modification by their environments. Accordingly, the assumption that no modification of these laws is to be looked for in environments, which have any striking difference from the environments for which the laws have been observed to hold, is very unsafe. The physical entities may be modified in very essential ways, so far as these laws are concerned. It is even possible that they may be developed into individualities of more fundamental types, with wider embodiment of envisagement. Such envisagement might reach to the attainment of the poising of alternative values with exercise of choice lying outside the physical laws, and expressible only in terms of purpose. Apart from such remote possibilities, it remains an immediate deduction that an individual entity, whose own life-history is a part within the life-history of some larger, deeper, more complete pattern, is liable to have aspects of that larger pattern dominating its own being, and to experience modifications of that larger pattern reflected in itself as modifications of its own being. This is the theory of organic mechanism.

According to this theory the evolution of laws of nature

is concurrent with the evolution of enduring pattern. For the general state of the universe, as it now is, partly determines the very essences of the entities whose modes of functioning these laws express. The general principle is that in a new environment there is an evolution of the old entities into new forms.

This rapid outline of a thoroughgoing organic theory of nature enables us to understand the chief requisites of the doctrine of evolution. The main work, proceeding during this pause at the end of the nineteenth century, was the absorption of this doctrine as guiding the methodology of all branches of science. By a blindness which is almost judicial as being a penalty affixed to hasty, superficial thinking, many religious thinkers opposed the new doctrine; although, in truth, a thoroughgoing evolutionary philosophy is inconsistent with materialism. The aboriginal stuff, or material, from which a materialistic philosophy starts is incapable of evolution. This material is in itself the ultimate substance. Evolution, on the materialistic theory, is reduced to the role of being another word for the description of the changes of the external relations between portions of matter. There is nothing to evolve, because one set of external relations is as good as any other set of external relations. There can merely be change, purposeless and unprogressive. But the whole point of the modern doctrine is the evolution of the complex organisms from antecedent states of less complex organisms. The doctrine thus cries aloud for a conception of organism as fundamental for nature. It also requires an underlying activity – a substantial activity – expressing itself in individual embodiments, and evolving in achievements of organism. The organism is a unit of emergent value, a real fusion of the characters of eternal objects, emerging for its own sake.

Thus in the process of analysing the character of nature in itself, we find that the emergence of organisms depends on a selective activity which is akin to purpose. The point is that the enduring organisms are now the outcome of evolution; and that, beyond these organisms, there is nothing else that endures. On the materialistic theory, there is material – such as matter or electricity – which endures. On the organic theory, the only endurances are structures of activity, and the structures are evolved.

Enduring things are thus the outcome of a temporal process; whereas eternal things are the elements required for the very being of the process. We can give a precise definition of endurance in this way: Let an event *A* be pervaded by an enduring structural pattern. Then *A* can be exhaustively subdivided into a temporal succession of events. Let *B* be any part of *A* which is obtained by picking out any one of the events belonging to a series which thus subdivides *A*. Then the enduring pattern is a pattern of aspects within the complete pattern prehended into the unity of *A*, and it is also a pattern within the complete pattern prehended into the unity of any temporal slice of *A*, such as *B*. For example, a molecule is a pattern exhibited in an event of one minute, and of any second of that minute. It is obvious that such an enduring pattern may be of more, or of less, importance. It may express some slight fact connecting the underlying activities thus individualized; or it may express some very close connection. If the pattern which endures is merely derived from the direct aspects of the external environment, mirrored in the standpoints of the various parts, then the endurance is an extrinsic fact of slight importance. But if the enduring pattern is wholly derived from the direct aspects of the various temporal sections of the event in question, then the endurance is an important intrinsic fact. It expresses a certain unity of character uniting the underlying individualized activities. There is then an enduring object with a certain unity for itself and for the rest of nature. Let us use the term physical endurance to express endurance of this type. Then physical endurance is the process of continuously inheriting a certain identity of character transmitted throughout a historical route of events. This character belongs to the whole route, and to every event of the route. This is the exact property of material. If it has existed for ten minutes, it has existed during every minute of the ten minutes, and during every second of every minute. Only if you take material to be fundamental, this property of endurance is an arbitrary fact at the base of the order of nature; but if you take organism to be fundamental, this property is the result of evolution.

It looks at first sight, as if a physical object, with its process of inheritance from itself, were independent of the environment. But such a conclusion is not justified. For let *B* and *C* be two successive slices in the life of such an object, such that *C* succeeds *B*. Then the enduring pattern in *C* is inherited from *B*, and from other analogous antecedent parts of its life. It is transmitted through *B* to *C*. But what is transmitted to *C* is the complete pattern of aspects derived from such events as *B*. These complete patterns include the influence of the environment on *B*, and on the other antecedent parts of the life of the object. Thus the complete aspects of the antecedent life are inherited as the partial pattern which endures throughout all the various periods of the life. Thus a favourable environment is essential to the maintenance of a physical object.

Nature, as we know it, comprises enormous permanences. There are the permanences of ordinary matter. The molecules within the oldest rocks known to geologists may have existed unchanged for over a thousand million years, not only unchanged in themselves, but unchanged in their relative dispositions to each other. In that length of time the number of pulsations of a molecule vibrating with the frequency of yellow sodium light would be about $16{\cdot}3 \times 10^{23}$ $= 163{,}000 \times (10^6)^3$. Until recently, an atom was apparently indestructible. We know better now. But the indestructible atom has been succeeded by the apparently indestructible electron and the indestructible proton.

Another fact to be explained is the great similarity of these practically indestructible objects. All electrons are very similar to each other. We need not outrun the evidence, and say that they are identical; but our powers of observation cannot detect any differences. Analogously, all hydrogen nuclei are alike. Also we note the great numbers of these analogous objects. There are throngs of them. It seems as though a certain similarity were a favourable condition for existence. Common sense also suggests this conclusion. If organisms are to survive, they must work together.

Accordingly, the key to the mechanism of evolution is the necessity for the evolution of a favourable environment, conjointly with the evolution of any specific type of enduring organisms of great permanence. Any physical object which by its influence deteriorates its environment, commits suicide.

One of the simplest ways of evolving a favourable environment concurrently with the development of the individual organism, is that the influence of each organism on the environment should be favourable to the endurance of other organisms of the same type. Further, if the organism also favours the development of other organisms of the same type, you have then obtained a mechanism of evolution adapted to produce the observed state of large multitudes of analogous entities, with high powers of endurance. For the environment automatically develops with the species, and the species with the environment.

The first question to ask is, whether there is any direct evidence for such a mechanism for the evolution of enduring organisms. In surveying nature, we must remember that there are not only basic organisms whose ingredients are merely aspects of eternal objects. There are also organisms of organisms. Suppose for the moment and for the sake of simplicity, we assume, without any evidence, that electrons and hydrogen nuclei are such basic organisms. Then the atoms, and the molecules, are organisms of a higher type, which also represent a compact definite organic unity. But when we come to the larger aggregations of matter, the organic unity fades into the background. It appears to be but faint and elementary. It is there; but the pattern is vague and indecisive. It is a mere aggregation of effects. When we come to living beings, the definiteness of pattern is recovered, and the organic character again rises into prominence. Accordingly, the characteristic laws of inorganic matter are mainly the statistical averages resulting from confused aggregates. So far are they from throwing light on the ultimate nature of things, that they blur and obliterate the individual characters of the individual organisms. If we wish to throw light upon the facts relating to organisms, we must study either the individual molecules and electrons, or the individual living beings. In between we find comparative confusion. Now the difficulty of studying the individual molecule is that we know so little about its life-history. We cannot keep an individual under continuous observation. In general, we deal with them in large aggregates. So far as individuals are concerned, sometimes with difficulty a great experimenter throws, so to speak, a flash light on one of them, and just observes one type of instantaneous effect. Accordingly, the history of the functioning of individual molecules, or electrons, is largely hidden from us.

But in the case of living beings, we can trace the history of individuals. We now find exactly the mechanism which is here demanded. In the first place, there is the propagation of the species from members of the same species. There is also the careful provision of the favourable environment for the endurance of the family, the race, or the seed in the fruit.

It is evident, however, that I have explained the evolutionary mechanism in terms which are far too simple. We find associated species of living things, providing for each other a favourable environment. Thus just as the members of the same species mutually favour each other, so do members of associated species. We find the rudimentary fact of association in the existence of the two species, electrons and hydrogen nuclei. The simplicity of the dual association, and the apparent absence of competition from other antagonistic species accounts for the massive endurance which we find among them.

There are thus two sides to the machinery involved in the development of nature. On one side, there is a given environment with organisms adapting themselves to it. The scientific materialism of the epoch in question emphasized this aspect. From this point of view, there is a given amount of material, and only a limited number of organisms can take advantage of it. The givenness of the environment dominates everything. Accordingly, the last words of science appeared to be the struggle for existence, and natural selection. Darwin's own writings are for all time a model of refusal to go beyond the direct evidence, and of careful retention of every possible hypothesis. But those virtues were not so conspicuous in his followers, and still less in his camp-followers. The imagination of European sociologists and publicists was stained by exclusive attention to this aspect of conflicting interests. The idea prevailed that there was a peculiar strong-minded realism in discarding ethical considerations in the determination of the conduct of commercial and national interests.

The other side of the evolutionary machinery, the neglected side, is expressed by the word creativeness. The organisms can create their own environment. For this purpose, the single organism is almost helpless. The adequate forces require societies of co-operating organisms. But with such co-operation and in proportion to the effort put forward, the environment has a plasticity which alters the whole ethical aspect of evolution.

In the immediate past, and at present, a muddled state of mind is prevalent. The increased plasticity of the environment for mankind, resulting from the advances in scientific technology, is being construed in terms of habits of thought which find their justification in the theory of a fixed environment.

The riddle of the universe is not so simple. There is the aspect of permanence in which a given type of attainment is endlessly repeated for its own sake; and there is the aspect of transition to other things – it may be of higher worth, and it may be of lower worth. Also there are its aspects of struggle and of friendly help. But romantic ruthlessness is no nearer to real politics, than is romantic self-abnegation.

CHAPTER VII

RELATIVITY

In the previous lectures of this course we have considered the antecedent conditions which led up to the scientific movement, and have traced the progress of thought from the seventeenth to the nineteenth century. In the nineteenth century this history falls into three parts, so far as it is to be grouped around science. These divisions are, the contact between the romantic movement and science, the development of technology and physics in the earlier part of the century, and lastly the theory of evolution combined with the general advance of the biological sciences.

The dominating note of the whole period of three centuries is that the doctrine of materialism afforded an adequate basis for the concepts of science. It was practically unquestioned. When undulations were wanted, an ether was supplied, in order to perform the duties of an undulatory material. To show the full assumption thus involved, I have sketched in outline an alternative doctrine of an organic theory of nature. In the last lecture it was pointed out that the biological developments, the doctrine of evolution, the doctrine of energy, and the molecular theories were rapidly undermining the adequacy of the orthodox materialism. But until the close of the century no one drew that conclusion. Materialism reigned supreme.

The note of the present epoch is that so many complexities have developed regarding material, space, time, and energy, that the simple security of the old orthodox assumptions has vanished. It is obvious that they will not do as Newton left them, or even as Clerk Maxwell left them. There must be a reorganization. The new situation in the thought of today arises from the fact that scientific theory is outrunning common sense. The settlement as inherited by the eighteenth century was a triumph of organized common sense. It had got rid of medieval phantasies, and of Cartesian vortices. As a result it gave full rein to its anti-rationalistic tendencies derived from the historical revolt of the Reformation period. It grounded itself upon what every plain man could see with his own eyes, or with a microscope of moderate power. It measured the obvious things to be measured, and it generalized the obvious things to be generalized. For example, it generalized the ordinary notions of weight and massiveness. The eighteenth century opened with the quiet confidence that at last nonsense had been got rid of. Today we are at the opposite pole of thought. Heaven knows what seeming nonsense may not tomorrow be demonstrated truth. We have recaptured some of the tone of the early nineteenth century, only on a higher imaginative level.

The reason why we are on a higher imaginative level is not because we have finer imagination, but because we have better instruments. In science, the most important thing that has happened during the last forty years is the advance in instrumental design. This advance is partly due to a few men of genius such as Michelson and the German opticians. It is also due to the progress of technological processes of manufacture, particularly in the region of metallurgy. The designer has now at his disposal a variety of material of differing physical properties. He can thus depend upon obtaining the material he desires; and it can be ground to the shapes he desires, within very narrow limits of tolerance. These instruments have put thought on to a new level. A fresh instrument serves the same purpose as foreign travel; it shows things in unusual combinations. The gain is more than a mere addition; it is a transformation. The advance in experimental ingenuity is, perhaps, also due to the larger proportion of national ability which now flows into scientific pursuits. Anyhow, whatever be the cause, subtle and ingenious experiments have abounded within the last generation. The result is, that a great deal of information has been accumulated in regions of nature very far removed from the ordinary experience of mankind.

Two famous experiments, one devised by Galileo at the outset of the scientific movement, and the other by Michelson with the aid of his famous interferometer, first carried out in 1881, and repeated in 1887 and 1905, illustrate the assertions I have made. Galileo dropped heavy bodies from the top of the leaning tower of Pisa, and demonstrated that bodies of different weights, if released simultaneously, would reach the earth together. So far as experimental skill and delicacy of apparatus were concerned, this experiment could have been made at any time within the preceding five thousand years. The ideas involved merely concerned weight and speed of travel, ideas which are familiar in ordinary life. The whole set of ideas might have been familiar to the family of King Minos of Crete, as they dropped pebbles into the sea from high battlements rising from the shore. We cannot too carefully realize that science started with the organization of ordinary experiences. It was in this way that it coalesced so readily with the antirationalistic bias of the historical revolt. It was not asking for ultimate meanings. It confined itself to investigating the connections regulating the succession of obvious occurrences.

Michelson's experiment could not have been made earlier than it was. It required the general advance in technology, and Michelson's experimental genius. It concerns the determination of the earth's motion through the ether, and it assumes that light consists of waves of vibration advancing at a fixed rate through the ether in any direction. Also, of course, the earth is moving through the ether, and Michelson's apparatus is moving with the earth. In the centre of the apparatus a ray of light is divided so that one half-ray goes in one direction *along* the apparatus through a given distance, and is reflected back to the centre by a mirror in the apparatus. The other half-ray goes the same distance *across* the apparatus in a direction at right angles to the former ray, and it also is reflected back to the centre. These reunited rays are then reflected on to a screen in the apparatus. If precautions are taken, you will see interference bands; namely bands of blackness where the crests of the waves of one ray have filled up the troughs of the other rays, owing to a minute difference in the lengths of paths of the two half-rays, up to certain parts of the screens. These differences in length will be affected by the motion of the earth. For it is the lengths of the paths in the ether which count. Thus, since the apparatus is moving with the earth, the path of one half-ray will be disturbed by the motion in a different manner from the path of the other half-ray. Think of yourself as moving in a railway carriage, first along the train and then across the train; and mark out your paths on the railway track which in this analogy corresponds to the ether. Now the motion of the earth is very slow compared to that of light. Thus in the analogy you must think of the train almost at a standstill, and of yourself as moving very quickly.

In the experiment this effect of the earth's motion would affect the positions on the screen of the interference bands. Also if you turn the apparatus round, through a right-angle, the effect of the earth's motion on the two half-rays will be interchanged, and the positions of the interference bands would be shifted. We can calculate the small shift which should result owing to the earth's motion round the sun. Also to this effect, we have to add that due to the sun's motion through the ether. The delicacy of the instrument can be tested, and it can be proved that these effects of shifting are large enough to be observed by it. Now the point is, that nothing was observed. There was no shifting as you turned the instrument round.

The conclusion is either that the earth is always stationary in the ether, or that there is something wrong with the fundamental principles on which the interpretation of the experiment relies. It is obvious that, in this experiment, we are very far away from the thoughts and the games of the children of King Minos. The ideas of an ether, of waves in it, of interference, of the motion of the earth through the ether, and of Michelson's interferometer, are remote from ordinary experience. But remote as they are, they are simple and obvious compared to the accepted explanation of the nugatory result of the experiment.

The ground of the explanation is that the ideas of space and of time employed in science are too simple-minded, and must be modified. This conclusion is a direct challenge to common sense, because the earlier science had only refined upon the ordinary notions of ordinary people. Such a radical reorganization of ideas would not have been adopted, unless it had also been supported by many other observations which we need not enter upon. Some form of the relativity theory seems to be the simplest way of explaining a large number of facts which otherwise would require some *ad hoc* explanation. The theory, therefore, does not merely depend upon the experiments which led to its origination.

The central point of the explanation is that every instrument, such as Michelson's apparatus as used in the experiment, necessarily records the velocity of light as having one and the same definite speed relatively to it. I mean that an interferometer in a comet and an interferometer on the earth would necessarily bring out the velocity of light, relatively to themselves, as at the same value. This is an obvious paradox, since the light moves with a definite velocity through the ether. Accordingly two bodies, the earth and the comet, moving with unequal velocities through the ether, might be expected to have different velocities relatively to rays of light. For example, consider two cars on a road, moving at ten and twenty miles an hour respectively, and being passed by another car at fifty miles an hour. The rapid car will pass one of the two cars at the relative velocity of forty miles per hour, and the other at the rate of thirty miles per hour. The allegation as to light is that, if we substituted a ray of light for the rapid car, the velocity of the light along the roadway would be exactly the same as its velocity relatively to either of the two cars which it overtakes. The velocity of light is immensely large, being about three hundred thousand kilometres per second. We must have notions as to space and time such that just this velocity has this peculiar character. It follows that all our notions of relative velocity must be recast. But these notions are the immediate outcome of our habitual notions as to space and time. So we come back to the position, that there has been something overlooked in the current expositions of what we mean by space and of what we mean by time.

Now our habitual fundamental assumption is that there is a unique meaning to be given to space and a unique meaning to be given to time, so that whatever meaning is given to spatial relations in respect to the instrument on the earth, the same meaning must be given to them in respect to the instrument on the comet, and the same meaning for an instrument at rest in the ether. In the theory of relativity, this is denied. As far as concerns space, there is no difficulty in agreeing, if you think of the obvious facts of relative motion. But even here the change in meaning has to go further than would be sanctioned by common sense. Also the same demand is made for time; so that the relative dating of events and the lapses of time between them are to be reckoned as different for the instrument on the earth, for the instrument in the comet, and for the instrument at rest in the ether. This is a greater strain on our credulity. We need not probe the question further than the conclusion that for the earth and for the comet spatiality and temporality are each to have different meanings amid different conditions, such as those presented by the earth and the comet. Accordingly velocity has different meanings for the two bodies. Thus the modern scientific assumption is that if anything has the speed of light by reference to any one meaning of space and time, then it has the same speed according to any other meaning of space and time.

This is a heavy blow at the classical scientific materialism, which presupposes a definite present instant at which all matter is simultaneously real. In the modern theory there is no such unique present instant. You can find a meaning for the notion of the simultaneous instant throughout all nature, but it will be a different meaning for different notions of temporality.

There has been a tendency to give an extreme subjectivist interpretation to this new doctrine. I mean that the relativity of space and time has been construed as though it were dependent on the choice of the observer. It is perfectly legitimate to bring in the observer, if he facilitates explanations. But it is the observer's body that we want, and not his mind. Even this body is only useful as an example of a very familiar form of apparatus. On the whole, it is better to concentrate attention on Michelson's interferometer, and to leave Michelson's body and Michelson's mind out of the picture. The question is, why did the interferometer have black bands on its screen, and why did not these bands slightly shift as the instrument turned. The new relativity associates space and time with an intimacy not hitherto contemplated; and presupposes that their separation in concrete fact can be achieved by alternative modes of abstraction, yielding alternative meanings. But each mode of abstraction is directing attention to something which is in nature; and thereby is isolating it for the purpose of contemplation. The fact relevant to experiment, is the relevance of the interferometer to just one among the many alternative systems of these spatio-temporal relations which hold between natural entities.

What we must now ask of philosophy is to give us an interpretation of the status in nature of space and time, so that the possibility of alternative meanings is preserved. These lectures are not suited for the elaboration of details; but there is no difficulty in pointing out where to look for the origin of the discrimination between space and time. I am presupposing the organic theory of nature, which I have outlined as a basis for a thoroughgoing objectivism.

An event is the grasping into unity of a pattern of aspects. The effectiveness of an event beyond itself arises from the aspects of itself which go to form the prehended unities of other events. Except for the systematic aspects of geometrical shape, this effectiveness is trivial, if the mirrored pattern attaches merely to the event as one whole. If the pattern endures throughout the successive parts of the event, and also exhibits itself in the whole, so that the event is the life-history of the pattern, then in virtue of that enduring pattern the event gains in external effectiveness. For its own effectiveness is re-enforced by the analogous aspects of all its successive parts. The event constitutes a patterned value with a permanence inherent throughout its own parts; and by reason of this inherent endurance the event is important for the modification of its environment.

It is in this endurance of pattern that time differentiates itself from space. The pattern is spatially *now*; and this temporal determination constitutes its relation to each partial event. For it is reproduced in this temporal succession of these spatial parts of its own life. I mean that this particular rule of temporal order allows the pattern to be reproduced in each temporal slice of its history. So to speak, each enduring object discovers in nature and requires from nature a principle discriminating space from time. Apart from the fact of an enduring pattern this principle might be there, but it would be latent and trivial. Thus the importance of space as against time, and of time as against space, has developed with the development of enduring organisms. Enduring objects are significant of a differentiation of space from time in respect to the patterns ingredient within events; and conversely the differentiation of space from time in the patterns ingredient within events expresses the patience of the community of events for enduring objects. There might be the community without objects, but there could not be the enduring objects without the community with its pecular patience for them.

It is very necessary that this point should not be misunderstood. Endurance means that a pattern which is exhibited in the prehension of one event is also exhibited in the prehension of those of its parts which are discriminated by a certain rule. It is not true that any part of the whole event will yield the same pattern as does the whole. For example, consider the total bodily pattern exhibited in the life of a human body during one minute. One of the thumbs during the same minute is part of the whole bodily event. But the pattern of this part is the pattern of the thumb, and is not the pattern of the whole body. Thus endurance requires a definite rule for obtaining the parts. In the above example, we know at once what the rule is: You must take the life of the whole body during any portion of that same minute; for example, during a second or a tenth of a second. In other words, the meaning of endurance presupposes a meaning for the lapse of time within the spatio-temporal continuum.

The question now arises whether all enduring objects discover the same principle of differentiation of space from time; or even whether at different stages of its own life-history one object may not vary in its spatio-temporal discrimination. Up till a few years ago, everyone unhesitatingly assumed that there was only one such principle to be discovered. Accordingly, in dealing with one object, time would have exactly the same meaning in reference to endurance as in dealing with the endurance of another object. It would also follow then that spatial relations would have one unique meaning. But now it seems that the observed effectiveness of objects can only be explained by assuming that objects in a state of motion relatively to each other are utilizing, for their endurance, meanings of space and of time which are not identical from one object to another. Every enduring object is to be conceived as at rest in its own proper space, and in motion throughout any space defined in a way which is not that inherent in its peculiar endurance. If two objects are mutually at rest, they are utilizing the same meanings of space and of time for the purposes of expressing their endurance; if in relative motion, the spaces and times differ. It follows that, if we can conceive a body at one stage of its life-history as in motion relatively to itself at another stage, then the body at these two stages is utilizing diverse meanings of space, and correlatively diverse meanings of time.

In an organic philosophy of nature there is nothing to decide between the old hypothesis of the uniqueness of the time discrimination and the new hypothesis of its multiplicity. It is purely a matter for evidence drawn from observations.[1]

In an earlier lecture, I said that an event had contemporaries. It is an interesting question whether, on the new hypothesis, such a statement can be made without the qualification of a reference to a definite space-time system. It is possible to do so, in the sense that in *some* time-system or other the two events are simultaneous. In other time-systems the two contemporary events will not be simultaneous, though they may overlap. Analogously one event will precede another without qualification, if in *every* time-system this precedence occurs. It is evident that if we start from a given event *A*, other events in general are divided into two sets, namely, those which without qualification are contemporaneous with *A* and those which either precede or succeed *A*. But there will be a set left over, namely, those events which bound the two sets. There we have a critical case. You will remember that we have a critical velocity to account for, namely the theoretical velocity of light *in vacuo*.[2] Also you will remember that the utilization of different spatio-temporal systems means the relative motion of objects. When we analyse this critical

[1] Cf. my *Principles of Natural Knowledge*, Sec. 52:3.

[2] This not the velocity of light in a gravitational field or in a medium of molecules and electrons.

relation of a special set of events to any given event *A*, we find the explanation of the critical velocity which we require. I am suppressing all details. It is evident that exactness of statement must be introduced by the introduction of points, and lines, and instants. Also that the origin of geometry requires discussion; for example, the measurement of lengths, the straightness of lines, and the flatness of planes, and perpendicularity. I have endeavoured to carry out these investigations in some earlier books, under the heading of the theory of extensive abstraction; but they are too technical for the present occasion.

If there be no one definite meaning to the geometrical relations of distance, it is evident that the law of gravitation needs restatement. For the formula expressing that law is that two particles attract each other in proportion to the product of their masses and the inverse square of their distances. This enunciation tacitly assumes that there is one definite meaning to be ascribed to the instant at which the attraction is considered, and also one definite meaning to be ascribed to distance. But distance is a purely spatial notion, so that in the new doctrine, there are an indefinite number of such meanings according to the space-time system which you adopt. If the two particles are relatively at rest, then we might be content with the space-time systems which they are both utilizing. Unfortunately this suggestion gives no hint as to procedure when they are not mutually at rest. It is, therefore, necessary to reformulate the law in a way which does not presuppose any particular space-time system. Einstein has done this. Naturally the result is more complicated. He introduced into mathematical physics certain methods of pure mathematics which render the formulae independent of the particular systems of measurement adopted. The new formula introduces various small effects which are absent in Newton's law. But for the major effects Newton's law and Einstein's law agree. Now these extra effects of Einstein's law serve to explain irregularities of the planet Mercury's orbit which by Newton's law were inexplicable. This is a strong confirmation of the new theory. Curiously enough, there is more than one alternative formula, based on the new theory of multiple space-time systems, having the property of embodying Newton's law and in addition of explaining the peculiarities of Mercury's motion. The only method of selection between them is to wait for experimental evidence respecting those effects on which the formulae differ. Nature is probably quite indifferent to the aesthetic preferences of mathematicians.

It only remains to add that Einstein would probably reject the theory of multiple space-time systems which I have been expounding to you. He would interpret his formula in terms of contortions in space-time which alter the invariance theory for measure properties, and of the proper times of each historical route. His mode of statement has the greater mathematical simplicity, and only allows of one law of gravitation, excluding the alternatives. But, for myself, I cannot reconcile it with the given facts of our experience as to simultaneity, and spatial arrangement. There are also other difficulties of a more abstract character.

The theory of the relationship between events at which we have now arrived is based first upon the doctrine that the relatednesses of an event are all internal relations, so far as concerns that event, though not necessarily so far as concerns the other *relata*. For example, the eternal objects, thus involved, are externally related to events. This internal relatedness is the reason why an event can be found only just where it is and how it is – that is to say, in just one definite set of relationships. For each relationship enters into the essence of the event; so that, apart from that relationship, the event would not be itself. This is what is meant by the very notion of internal relations. It has been usual, indeed, universal, to hold that spatio-temporal relationships are external. This doctrine is what is here denied.

The conception of internal relatedness involves the

analysis of the event into two factors, one the underlying substantial activity of individualization, and the other the complex of aspects – that is to say, the complex of relatednesses as entering into the essence of the given event – which are unified by this individualized activity. In other words, the concept of internal relations requires the concept of substance as the activity synthesizing the relationships into its emergent character. The event is what it is, by reason of the unification in itself of a multiplicity of relationships. The general scheme of these mutual relationships is an abstraction which presupposes each event as an independent entity, which it is not, and asks what remnant of these formulative relationships is then left in the guise of external relationships. The scheme of relationships as thus impartially expressed becomes the scheme of a complex of events variously related as wholes to parts and as joint parts within some one whole. Even here, the internal relationship forces itself on our attention; for the part evidently is constitutive of the whole. Also an isolated event which has lost its status in any complex of events is equally excluded by the very nature of an event. So the whole is evidently constitutive of the part. Thus the internal character of the relationship really shows through this impartial scheme of abstract external relations.

But this exhibition of the actual universe as extensive and divisible has left out the distinction between space and time. It has in fact left out the process of realization, which is the adjustment of the synthetic activities by virtue of which the various events become their realized selves. This adjustment is thus the adjustment of the underlying active substances whereby these substances exhibit themselves as the individualizations or modes of Spinoza's one substance. This adjustment is what introduces temporal process.

Thus, in some sense, time, in its character of the adjustment of the process of synthetic realization, extends beyond the spatio-temporal continuum of nature.[3] There is no

[3] Cf. my *Concept of Nature*, Ch. 111.

necessity that temporal process, in this sense, should be constituted by one single series of linear succession. Accordingly, in order to satisfy the present demands of scientific hypothesis, we introduce the metaphysical hypothesis that this is not the case. We do assume (basing ourselves upon direct observation), however, that temporal process of realization can be analysed into a group of linear serial processes. Each of these linear series is a space-time system. In support of this assumption of definite serial processes, we appeal : (1) to the immediate presentation through the senses of an extended universe beyond ourselves and simultaneous with ourselves, (2) to the intellectual apprehension of a meaning to the question which asks what is now immediately happening in regions beyond the cognizance of our senses, (3) to the analysis of what is involved in the endurance of emergent objects. This endurance of objects involves the display of a pattern as now realized. This display is the display of a pattern as inherent in an event, but also as exhibiting a temporal slice of nature as lending aspects to eternal objects (or, equally, of eternal objects as lending aspects to events). The pattern is spatialized in a whole duration for the benefit of the event into whose essence the pattern enters. The event is part of the duration, i.e., is part of what is exhibited in the aspects inherent in itself; and conversely the duration is the whole of nature simultaneous with the event, in that sense of simultaneity. Thus an event in realizing itself displays a pattern, and this pattern requires a definite duration determined by a definite meaning of simultaneity. Each such meaning of simultaneity relates the pattern as thus displayed to one definite space-time system. The actuality of the space-time systems is constituted by the realization of pattern; but it is inherent in the general scheme of events as constituting its patience for the temporal process of realization.

Notice that the pattern requires a duration involving a definite lapse of time, and not merely an instantaneous moment. Such a moment is more abstract, in that it merely denotes a certain relation of contiguity between the concrete events. Thus a duration is spatialized; and by 'spatialized' is meant that the duration is the field for the realiized pattern constituting the character of the event. A duration, as the field of the pattern realized in the actualization of one of its contained events, is an epoch, i.e., an arrest. Endurance is the repetition of the pattern in successive events. Thus endurance requires a succession of durations, each exhibiting the pattern. In this account 'time' has been separated from 'extension' and from the 'divisibility' which arises from the character of spatio-temporal of extension. Accordingly we must not proceed to conceive time as another form of extensiveness. Time is sheer succession of epochal durations. But the entities which succeed each other in this account are durations. The duration is that which is required for the realization of a pattern in the given event. Thus the divisibility and extensiveness is within the given duration. The epochal duration is not realized via its successive divisible parts, but is given with its parts. In this way, the objection which Zeno might make to the joint validity of two passages from Kant's *Critique of Pure Reason* is met by abandoning the earlier of the two passages. I refer to passages from the section 'Of the Axioms of Intuition'; the earlier from the subsection on 'Extensive Quantity', and the latter from the subsection on 'Intensive Quantity' where considerations respecting quantity in general, extensive and intensive, are summed up. The earlier passage runs thus[4]:

> I call an extensive quantity that in which the representation of the whole is rendered possible by the representation of its parts, *and therefore necessarily preceded by it.*[5] I cannot represent to myself any line, however small it may be, without drawing it in thought, that is, without producing all its parts one after the other, start-

[4] Max Müller's translation.

[5] Italics mine, and also in the second passage.

ing from a given point, and thus, first of all, drawing its intuition. The same applies to every, even the smallest portion of time. I can only think in it the successive progress from one moment to another, thus producing in the end, by all the portions of time, and their addition, a definite quantity of time.

The second passage runs thus:

This peculiar property of quantities that no part of them is the smallest possible part (no part indivisible) is called continuity. Time and space are quanta continua, because there is no part of them that is not enclosed between limits (points and moments), *no part that is not itself again a space or a time. Space consists of spaces only, time of times. Points and moments are only limits,* mere places of limitation, and as places *presupposing always* those intuitions which they are meant to limit or to determine. Mere places or parts that might be given before space, or time, could never be compounded into space or time.

I am in complete agreement with the second extract if 'time and space' is the extensive continuum; but it is inconsistent with its predecessor. For Zeno would object that a vicious infinite regress is involved. Every part of time involves some small part of itself, and so on. Also this series regresses backwards ultimately to nothing; since the initial moment is without duration and merely marks the relation of contiguity to an earlier time. Thus time is impossible, if the two extracts are both adhered to. I accept the latter, and reject the earlier, passage. Realization is the becoming of time in the field of extension. Extension is the complex of events, *qua* their potentialities. In realization the potentiality becomes actuality. But the potential pattern requires a duration; and the duration must be exhibited as an epochal whole, by the realization of the pattern. Thus time is the succession of elements in themselves divisible and contiguous. A duration, in becoming temporal, thereby incurs realization in respect to some enduring object. Temporalization is realization. Temporalization is not another continuous process. It is an atomic succession. Thus time is atomic (i.e. epochal), though what is temporalized is divisible. This doctrine follows from the doctrine of events, and of the nature of enduring objects. In the next chapter we must consider its relevance to the quantum theory of recent science.

It is to be noted that this doctrine of the epochal character of time does not depend on the modern doctrine of relativity, and holds equally – and indeed, more simply – if this doctrine be abandoned. It does depend on the analysis of the intrinsic character of an event, considered as the most concrete finite entity.

In reviewing this argument, note first that the second quotation from Kant, on which it is based, does not depend on any peculiar Kantian doctrine. The latter of the two is in agreement with Plato as against Aristotle.[6] In the second place, the argument assumes that Zeno understated his argument. He should have urged it against the current notion of time in itself, and not against motion, which involves relations between time and space. For, what becomes has duration. But no duration can become until a smaller duration (part of the former) has antecedently come into being (Kant's earlier statement). The same argument applies to this smaller duration, and so on. Also the infinite regress of these durations converges to nothing – and even to the Aristotelian view there is no first moment. Accordingly time would be an irrational notion. Thirdly, in the epochal theory Zeno's difficulty is met by conceiving temporalization as the realization of a complete organism. This organism is an event holding in its essence its spatio-temporal relationships (both within itself, and beyond itself) throughout the spatio-temporal continuum.

[6] Cf. *Euclid in Greek,* by Sir T. L. Heath, Camb. Univ. Press, in a note on Points.

CHAPTER VIII

THE QUANTUM THEORY

The theory of relativity has justly excited a great amount of public attention. But, for all its importance, it has not been the topic which has chiefly absorbed the recent interest of physicists. Without question that position is held by the quantum theory. The point of interest in this theory is that, according to it, some effects which appear essentially capable of gradual increase or gradual diminution are in reality to be increased or decreased only by certain definite jumps. It is as though you could walk at three miles per hour or at four miles per hour, but not at three and a half miles per hour.

The effects in question are concerned with the radiation of light from a molecule which has been excited by some collision. Light consists of waves of vibration in the electromagnetic field. After a complete wave has passed a given point everything at that point is restored to its original state and is ready for the next wave which follows on. Picture to yourselves the waves on the ocean, and reckon from crest to crest of successive waves. The number of waves which pass a given point in one second is called the frequency of that system of waves. A system of light-waves of definite frequency corresponds to a definite colour in the spectrum. Now a molecule, when excited, vibrates with a certain number of definite frequencies. In other words, there are a definite set of modes of vibration of the molecule, and each mode of vibration has one definite frequency. Each mode of vibration can stir up in the electromagnetic field waves of its own frequency. These waves carry away the energy of the vibration; so that finally (when such waves are in being) the molecule loses the energy of its excitement and the waves cease. Thus a molecule can radiate light of certain definite colours, that is to say, of certain definite frequencies.

You would think that each mode of vibration could be excited to any intensity, so that the energy carried away by light of that frequency could be of any amount. But this is not the case. There appear to be certain minimum amounts of energy which cannot be subdivided. The case is analogous to that of a citizen of the United States who, in paying his debts in the currency of his country, cannot subdivide a cent so as to correspond to some minute subdivision of the goods obtained. The cent corresponds to the minimum quantity of the light energy, and the goods obtained correspond to the energy of the exciting cause. This exciting cause is either strong enough to procure the emission of one cent of energy, or fails to procure the emission of any energy whatsoever. In any case the molecule will only emit an integral number of cents of energy. There is a further peculiarity which we can illustrate by bringing an Englishman on to the scene. He pays his debts in English currency, and his smallest unit is a farthing which differs in value from the cent. The farthing is in fact about half a cent, to a very rough approximation. In the molecule, different modes of vibration have different frequencies. Compare each mode to a nation. One mode corresponds to the United States, and another mode corresponds to England. One mode can only radiate energy in an integral number of cents, so that a cent of energy is the least it can pay out; whereas the other mode can only radiate its energy in an integral number of farthings, so that a farthing of energy is the least that it can pay out. Also a rule can be found to tell us the relative value of the cent of energy of one mode to the farthing of energy of another mode. The rule is childishly simple: Each smallest coin of energy has a value in strict proportion to the frequency belonging to that mode. By this rule, and comparing farthings with cents, the frequency of an American would be about twice that of an Englishman. In other words, the American would do about twice as many things in a second as an Englishman. I must leave you to judge whether this corresponds to the reputed characters of the two nations. Also I suggest that there are merits attaching to both ends of the solar spectrum. Sometimes you want red light and sometimes violet light.

There has been, I hope, no great difficulty in comprehending what the quantum theory asserts about molecules. The perplexity arises from the effort to fit the theory into the current scientific picture of what is going on in the molecule or atom.

It has been the basis of the materialistic theory, that the happenings of nature are to be explained in terms of the locomotion of material. In accordance with this principle, the waves of light were explained in terms of the locomotion of a material ether, and the internal happenings of a molecule are now explained in terms of the locomotion of separate material parts. In respect to waves of light, the material ether has retreated to an indeterminate position in the background, and is rarely talked about. But the principle is unquestioned as regards its application to the atom. For example a neutral hydrogen atom is assumed to consist of at least two lumps of material; one lump is the nucleus consisting of a material called positive electricity, and the other is a single electron which is negative electricity. The nucleus shows signs of being complex, and of being subdivisible into smaller lumps, some of positive electricity and others electronic. The assumption is, that whatever vibration takes place in the atom is to be attributed to the vibratory locomotion of some bit of material, detachable from the remainder. The difficulty with the quantum theory is that, on this hypothesis, we have to picture the atom as providing a limited number of definite grooves, which are the sole tracks along which vibration can take place, whereas the classical scientific picture provides none of these grooves. The quantum theory wants trolley-cars with a limited number of routes, and the scientific picture provides horses galloping over prairies. The result is that the physical doctrine of the atom has got into a state which is strongly suggestive of the epicycles of astronomy before Copernicus.

On the organic theory of nature there are two sorts of vibrations which radically differ from each other. There is vibratory locomotion, and there is vibratory organic deformation; and the conditions for the two types of changes are of a different character. In other words, there is vibratory locomotion of a given pattern as one whole, and there is vibratory change of pattern.

A complete organism in the organic theory is what corresponds to a bit of material on the materialistic theory. There will be primary genus, comprising a number of species of organisms, such that each primary organism, belonging to a species of the primary genus, is not decomposable into subordinate organisms. I will call any organism of the primary genus a primate. There may be different species of primates.

It must be kept in mind that we are dealing with the abstractions of physics. Accordingly, we are not thinking of what a primate is in itself, as a pattern arising from the prehension of the concrete aspects; nor are we thinking of what a primate is for its environment, in respect to its concrete aspects prehended therein. We are thinking of these various aspects merely in so far as their efforts on patterns and on locomotion are expressible in spatio-temporal terms. Accordingly, in the language of physics, the aspects of a primate are merely its contributions to the electromagnetic field. This is in fact exactly what we know of electrons and protons. An electron for us is merely the pattern of its aspects in its environment, so far as those aspects are relevant to the electromagnetic field.

Now in discussing the theory of relativity, we saw that the relative motion of two primates means simply that their organic patterns are utilizing diverse space-time systems. If two primates do not continue either mutually at rest, or mutually in uniform relative motion, at least one of them is changing its intrinsic space-time system. The laws of motion express the conditions under which these changes of space-time systems are effected. The conditions for vibratory locomotion are founded upon these general laws of motion.

But it is possible that certain species of primates are apt to go to pieces under conditions which lead them to effect changes of space-time systems. Such species would only experience a long range of endurance, if they had succeeded in forming a favourable association among primates of different species, such that in this association the tendency to collapse is neutralized by the environment of the association. We can imagine the atomic nucleus as composed of a large number of primates of differing species, and perhaps with many primates of the same species, the whole association being such as to favour stability. An example of such an association is afforded by the association of a positive nucleus with negative electrons to obtain a neutral atom. The neutral atom is thereby shielded from any electric field which would otherwise produce changes in the space-time system of the atom.

The requirements of physics now suggest an idea which is very consonant with the organic philosophical theory. I put it in the form of a question : Has our organic theory of endurance been tainted by the materialistic theory in so far as it assumes without question that endurance must mean undifferentiated sameness throughout the life-history concerned? Perhaps you noticed that (in a previous chapter) I used the word 'reiteration' as a synonym of 'endurance'. It obviously is not quite synonymous in its meaning; and now I want to suggest that reiteration where it differs from endurance is more nearly what the organic theory requires. The difference is very analogous to that between the Galileans and the Aristotelians: Aristotle said 'rest' where

Galileo added 'or uniform motion in a straight line'. Thus in the organic theory, a pattern need not endure in undifferentiated sameness through time. The pattern may be essentially one of aesthetic contrasts requiring a lapse of time for its unfolding. A tune is an example of such a pattern. Thus the endurance of the pattern now means the reiteration of its succession of contrasts. This is obviously the most general notion of endurance on the organic theory, and 'reiteration' is perhaps the word which expresses it with most directness. But when we translate this notion into the abstractions of physics, it at once becomes the technical notion of 'vibration'. This vibration is not the vibratory locomotion : it is the vibration of organic deformation. There are certain indications in modern physics that for the role of corpuscular organisms at the base of the physical field, we require vibratory entities. Such corpuscles would be the corpuscles detected as expelled from the nuclei of atoms, which then dissolve into waves of light. We may conjecture that such a corpuscular body has no great stability of endurance, when in isolation. Accordingly, an unfavourable environment leading to rapid changes in its proper space-time system, that is to say, an environment jolting it into violent accelerations, causes the corpuscles to go to pieces and dissolve into light-waves of the same period of vibration.

A proton, and perhaps an electron, would be an association of such primates, superposed on each other, with their frequencies and spatial dimensions so arranged as to promote the stability of the complex organism, when jolted into acceleration of locomotion. The conditions for stability would give the associations of periods possible for protons. The expulsion of a primate would come from a jolt which leads the proton either to settle down into an alternative association, or to generate a new primate by the aid of the energy received.

A primate must be associated with a definite frequency of vibratory organic deformation so that when it goes to pieces it dissolves into light-waves of the same frequency, which then carry off all its average energy. It is quite easy (as a particular hypothesis) to imagine stationary vibrations of the electromagnetic field of definite frequency, and directed radially to and from a centre, which, in accordance with the accepted electromagnetic laws, would consist of a vibratory spherical nucleus satisfying one set of conditions and a vibratory external field satisfying another set of conditions. This is an example of vibratory organic deformation. Further (on this particular hypothesis), there are two ways of determining the subsidiary conditions so as to satisfy the ordinary requirements of mathematical physics. The total energy, according to one of these ways, would satisfy the quantum condition; so that it consists of an integral number of units or cents, which are such that the cent of energy of any primate is proportional to its frequency. I have not worked out the condition for stability or for a stable association. I have mentioned the particular hypothesis by way of showing by example that the organic theory of nature affords possibilties for the reconsideration of ultimate physical laws, which are not open to the opposed materialistic theory.

In this particular hypothesis of vibratory primates, the Maxwellian equations are supposed to hold throughout all space, including the interior of a proton. They express the laws governing the vibratory production and absorption of energy. The whole process for each primate issues in a certain average energy characteristic of the primate, and proportional to its mass. In fact the energy is the mass. There are vibratory radial streams of energy, both without and within a primate. Within the primate, there are vibratory distributions of electric density. On the materialistic theory such density marks the presence of material : on the organic theory of vibration, it marks the vibratory production of energy. Such production is restricted to the interior of the primate.

All science must start with some assumptions as to the

ultimate analysis of the facts with which it deals. These assumptions are justified partly by their adherence to the types of occurrence of which we are directly conscious, and partly by their success in representing the observed facts with a certain generality, devoid of *ad hoc* suppositions. The general theory of the vibration of primates, which I have outlined, is merely given as an example of the sort of possibilities which the organic theory leaves open for physical science. The point is that it adds the possibility of organic deformation to that of mere locomotion. Light waves form one great example of organic deformation.

At any epoch the assumptions of a science are giving way, when they exhibit symptoms of the epicyclic state from which astronomy was rescued in the sixteenth century. Physical science is now exhibiting such symptoms. In order to reconsider its foundations, it must recur to a more concrete view of the character of real things, and must conceive its fundamental notions as abstractions derived from this direct intuition. It is in this way that it surveys the general possibilities of revision which are open to it.

The discontinuities introduced by the quantum theory require revision of physical concepts in order to meet them. In particular, it has been pointed out that some theory of discontinuous existence is required. What is asked from such a theory, is that an orbit of an electron can be regarded as a series of detached positions, and not as a continuous line.

The theory of a primate or a vibrating pattern given above, together with the distinction between temporality and extensiveness in the previous chapter, yields exactly this result. It will be remembered that the continuity of the complex of events arises from the relationships of extensiveness; whereas the temporality arises from the realization in a subject-event of a pattern which requires for its display that the whole of a duration be spatialized (i.e. arrested), as given by its aspects in the event. Thus realization proceeds via a succession of epochal durations; and the continuous transition, i.e. the organic deformation, is within the duration which is already given. The vibratory organic deformation is in fact the reiteration of the pattern. One complete period defines the duration required for the complete pattern. Thus the primate is realized atomically in a succession of durations, each duration to be measured from one maximum to another. Accordingly, so far as the primate as one enduring whole entity is to be taken account of, it is to be assigned to these durations successively. If it is considered as one thing, its orbit is to be diagrammatically exhibited by a series of detached dots. Thus the locomotion of the primate is discontinuous in space and time. If we go below the quanta of time which are the successive vibratory periods of the primate, we find a succession of vibratory electromagnetic fields, each stationary in the space-time of its own duration. Each of these fields exhibits a single complete period of the electromagnetic vibration which constitutes the primate. This vibration is not to be thought of as the becoming of reality; it is what the primate is in one of its discontinuous realizations. Also the successive durations in which the primate is realized are contiguous; it follows that the life-history of the primate can be exhibited as being the continuous development of occurrences in the electromagnetic field. But these occurrences enter into realization as whole atomic blocks, occupying definite periods of time.

There is no need to conceive that time is atomic in the sense that all patterns must be realized in the same successive durations. In the first place, even if the periods were the same in the case of two primates, the durations of realization may not be the same. In other words, the two primates may be out of phase. Also if the periods are different, the atomism of any one duration of primate is necessarily subdivided by the boundary moments of durations of the other primate.

The laws of the locomotion of primates express under

what conditions any primate will change its space-time system.

It is unnecessary to pursue this conception further. The justification of the concept of vibratory existence must be purely experimental. The point illustrated by this example is that the cosmological outlook, which is here adopted, is perfectly consistent with the demands for discontinuity which have been urged from the side of physics. Also if this concept of temporalization as a successive realization of epochal durations be adopted, the difficulty of Zeno is evaded. The particular form, which has been given here to this concept, is purely for that purpose of illustration and must necessarily require recasting before it can be adapted to the results of experimental physics.

CHAPTER IX

SCIENCE AND PHILOSOPHY

In the present lecture, it is my object to consider some reactions of science upon the stream of philosophic thought during the modern centuries with which we are concerned. I shall make no attempt to compress a history of modern philosophy within the limits of one lecture. We shall merely consider some contacts between science and philosophy, in so far as they lie within the scheme of thought which it is the purpose of these lectures to develop. For this reason the whole of the great German idealistic movement will be ignored, as being out of effective touch with its contemporary science so far as reciprocal modification of concepts is concerned. Kant, from whom this movement took its rise, was saturated with Newtonian physics, and with the ideas of the great French physicists – such as Clairaut,[1] for instance – who developed the Newtonian ideas. But the philosophers who developed the Kantian school of thought, or who transformed it into Hegelianism, either lacked Kant's background of scientific knowledge, or lacked his potentiality of becoming a great physicist if philosophy had not absorbed his main energies.

The origin of modern philosophy is analogous to that of science, and is contemporaneous. The general trend of its

[1] Cf. the curious evidence of Kant's scientific reading in the *Critique of Pure Reason, Transcendental Analytic, Second Analogy of Experience,* where he refers to the phenomenon of capillary action. This is an unnecessarily complex illustration; a book resting on a table would have equally well sufficed. But the subject had just been adequately treated for the first time by Clairaut in an appendix to his *Figure of the Earth*. Kant evidently had read this appendix, and his mind was full of it.

development was settled in the seventeenth century, partly at the hands of the same men who established the scientific principles. This settlement of purpose followed upon a transitional period dating from the fifteenth century. There was in fact a general movement of European mentality, which carried along with its stream, religion, science and philosophy. It may shortly be characterized as being the direct recurrence to the original sources of Greek inspiration on the part of men whose spiritual shape had been derived from inheritance from the Middle Ages. There was therefore no revival of Greek mentality. Epochs do not rise from the dead. The principles of aesthetics and of reason, which animated the Greek civilization, were reclothed in a modern mentality. Between the two there lay other religions, other systems of law, other anarchies, and other racial inheritances, dividing the living from the dead.

Philosophy is peculiarly sensitive to such differences. For, whereas you can make a replica of an ancient statue, there is no possible replica of an ancient state of mind. There can be no nearer approximation than that which a masquerade bears to real life. There may be understanding of the past, but there is a difference between the modern and the ancient reactions to the same stimuli.

In the particular case of philosophy, the distinction in tonality lies on the surface. Modern philosophy is tinged with subjectivism, as against the objective attitude of the ancients. The same change is to be seen in religion. In the early history of the Christian Church, the theological interest centred in discussions on the nature of God, the meaning of the Incarnation, and apocalyptic forecasts of the ultimate fate of the world. At the Reformation, the Church was torn asunder by dissension as to the individual experiences of believers in respect to justification. The individual subject of experience had been substituted for the total drama of all reality. Luther asked, 'How am I justified?'; modern philosophers have asked, 'How do I have knowledge?' The emphasis lies upon the subject of experience. This change of standpoint is the work of Christianity in its pastoral aspect of shepherding the company of believers. For century after century it insisted upon the infinite worth of the individual human soul. Accordingly, to the instinctive egotism of physical desires, it has superadded an instinctive feeling of justification for an egotism of intellectual outlook. Every human being is the natural guardian of his own importance. Without a doubt, this modern direction of attention emphasizes truths of the highest value. For example, in the field of practical life, it has abolished slavery, and has impressed upon the popular imagination the primary rights of mankind.

Descartes, in his *Discourse on Method*, and in his *Meditations*, discloses with great clearness the general conceptions which have since influenced modern philosophy. There is a subject receiving experience: in the *Discourse* this subject is always mentioned in the first person, that is to say, as being Descartes himself. Descartes starts with himself as being a mentality, which in virtue of its consciousness of its own inherent presentations of sense and of thought, is thereby conscious of its own existence as a unit entity. The subsequent history of philosophy revolves round the Cartesian formulation of the primary datum. The ancient world takes its stand upon the drama of the universe, the modern world upon the inward drama of the soul. Descartes, in his *Meditations*, expressly grounds the existence of this inward drama upon the possibility of error. There may be no correspondence with objective fact, and thus there must be a soul with activities whose reality is purely derivative from itself. For example, here is a quotation[2] from his second *Meditation*:

> But it will be said that these presentations are false, and that I am dreaming. Let it be so. At all events it is certain that I seem to see light, hear a noise, and feel heat; this cannot be false, and this is what in me is

[2] Quoted from Veitch's translation.

properly called perceiving (*sentire*), which is nothing else than thinking. From this I begin to know what I am with somewhat greater clearness and distinctness than heretofore.

Again in his third *Meditation*:

> . . .; for, as I before remarked, although the things which I perceive or imagine are perhaps nothing at all apart from me, I am nevertheless assured that those modes of consciousness which I call perceptions and imaginations, in as far only as they are modes of consciousness, exist in me.

The objectivism of the medieval and the ancient worlds passed over into science. Nature is there conceived as for itself, with its own mutual reactions. Under the recent influence of relativity, there has been a tendency towards subjectivist formulations. But, apart from this recent exception, nature, in scientific thought, has had its laws formulated without any reference to dependence on individual observers. There is, however, this difference between the older and the later attitudes towards science. The antirationalism of the moderns has checked any attempt to harmonize the ultimate concepts of science with ideas drawn from a more concrete survey of the whole of reality. The material, the space, the time, the various laws concerning the transition of material configurations, are taken as ultimate stubborn facts, not to be tampered with.

The effect of this antagonism to philosophy has been equally unfortunate both for philosophy and for science. In this lecture we are concerned with philosophy. Philosophers are rationalists. They are seeking to go behind stubborn and irreducible facts: they wish to explain in the light of universal principles the mutual reference between the various details entering into the flux of things. Also, they seek such principles as will eliminate mere arbitrariness; so that, whatever portion of fact is assumed or given, the existence of the remainder of things shall satisfy some demand of rationality. They demand meaning. In the words of Henry Sidgwick[3]:

> It is the primary aim of philosophy to unify completely, bring into clear coherence, all departments of rational thought, and this aim cannot be realized by any philosophy that leaves out of its view the important body of judgements and reasonings which form the subject matter of ethics.

Accordingly, the bias towards history on the part of the physical and social sciences with their refusal to rationalize below some ultimate mechanism, has pushed philosophy out of the effective currents of modern life. It has lost its proper role as a constant critic of partial formulations. It has retreated into the subjectivist sphere of mind, by reason of its expulsion by science from the objectivist sphere of matter. Thus the evolution of thought in the seventeenth century co-operated with the enhanced sense of individual personality derived from the Middle Ages. We see Descartes taking his stand upon his own ultimate mind, which his philosophy assures him of; and asking about its relations to the ultimate matter – exemplified, in the second *Meditation*, by the human body and a lump of wax – which his science assumes. There is Aaron's rod, and the magicians' serpents; and the only question for philosophy is, which swallows which; or whether, as Descartes thought, they all lived happily together. In this stream of thought are to be found Locke, Berkeley, Hume, Kant. Two great names lie outside this list, Spinoza and Liebniz. But there is a certain isolation of both of them in respect to their philosophical influence so far as science is concerned; as though they had strayed to extremes which lie outside the boundaries of safe philosophy, Spinoza by retaining

[3] Cf. Henry Sidgwick: *A Memoir*, Appendix I.

older ways of thought, and Leibniz by the novelty of his monads.

The history of philosophy runs curiously parallel to that of science. In the case of both, the seventeenth century set the stage for its two successors. But with the twentieth century a new act commences. It is an exaggeration to attribute a general change in a climate of thought to any one piece of writing, or to any one author. No doubt Descartes only expressed definitely and in decisive form what was already in the air of his period. Analogously, in attributing to William James the inauguration of a new stage in philosophy, we should be neglecting other influences of his time. But, admitting this, there still remains a certain fitness in contrasting his essay, *Does Consciousness Exist* published in 1904, with Descartes' *Discourse on Method*, published in 1637. James clears the stage of the old paraphernalia; or rather he alters its lighting. Take for example these two sentences from his essay:

> To deny plumply that 'consciousness' exists seems so absurd on the face of it – for undeniably 'thoughts' do exist – that I fear some readers will follow me no farther. Let me then immediately explain that I mean only to deny that the word stands for an entity, but to insist most emphatically that it does stand for a function.

The scientific materialism and the Cartesian ego were both challenged at the same moment, one by science and the other by philosophy, as represented by William James with his psychological antecedents; and the double challenge marks the end of a period which lasted for about two hundred and fifty years. Of course, 'matter' and 'consciousness' both express something so evident in ordinary experience that any philosophy must provide some things which answer to their respective meanings. But the point is that, in respect to both of them, the seventeenth-century settlement was infected with a presupposition which is now challenged. James denies that consciousness is an entity, but admits that it is a function. The discrimination between an entity and a function is therefore vital to the understanding of the challenge which James is advancing against the older modes of thought. In the essay in question, the character which James assigns to consciousness is fully discussed. But he does not unambiguously explain what he means by the notion of an entity, which he refuses to apply to consciousness. In the sentence which immediately follows the one which I have already quoted, he says:

> There is, I mean, no aboriginal stuff or quality of being, contrasted with that of which material objects are made, out of which our thoughts of them are made; but there is a function in experience which thoughts perform, and for the performance of which this quality of being is invoked. That function is *knowing*. 'Consciousness' is supposed necessary to explain the fact that things not only are, but get reported, are known.

Thus James is denying that consciousness is a 'stuff'.

The term 'entity', or even that of 'stuff', does not fully tell its own tale. The notion of 'entity' is so general that it may be taken to mean anything that can be thought about. You cannot think of mere nothing; and the something which is an object of thought may be called an entity. In this sense, a function is an entity. Obviously, this is not what James had in his mind.

In agreement with the organic theory of nature which I have been tentatively putting forward in these lectures, I shall for my own purposes construe James as denying exactly what Descartes asserts in his *Discourse* and his *Meditations*. Descartes discriminates two species of entities, matter and soul. The essence of matter is spatial extension; the essence of soul is its cogitation, in the full sense which Descartes assigns to the word *cogitare*, for example, in Section Fifty-three of Part I of his *Principles of Philosophy*, he enunciates:

> That of every substance there is one principal attribute, as thinking of the mind, extension of the body.

In the earlier, Fifty-first Section, Descartes states:

> By substance we can conceive nothing else than a thing which exists in such a way as to stand in need of nothing beyond itself in order to its existence.

Furthermore, later on, Descartes says:

> For example, because any substance which ceases to endure ceases also to exist, duration is not distinct from substance except in thought; . . .

Thus we conclude that, for Descartes, minds and bodies exist in such a way as to stand in need of nothing beyond themselves individually (God only excepted, as being the foundation of all things); that both minds and bodies endure, because without endurance they would cease to exist; that spatial extension is the essential attribute of bodies; and that cogitation is the essential attribute of minds.

It is difficult to praise too highly the genius exhibited by Descartes in the complete sections of his *Principles* which deal with these questions. It is worthy of the century in which he writes, and of the clearness of the French intellect. Descartes in his distinction between time and duration, and in his way of grounding time upon motion, and in his close relation between matter and extension, anticipates, as far as it was possible at his epoch, modern notions suggested by the doctrine of relativity, or by some aspects of Bergson's doctrine of the generation of things. But the fundamental principles are so set out as to presuppose independently existing substances with simple location in the community of temporal durations, and in the case of bodies, with simple location in the community of spatial extensions. Those principles lead straight to the theory of a materialistic, mechanistic nature, surveyed by cogitating minds. After the close of the seventeenth century, science took charge of the cogitating minds. Some schools of philosophy admitted an ultimate dualism; and the various idealistic schools claimed that nature was merely the chief example of the cogitations of minds. But all schools admitted the Cartesian analysis of the ultimate elements of nature. I am excluding Spinoza and Leibniz from these statements as to the main stream of modern philosophy, as derivative from Descartes; though of course they were influenced by him, and in their turn influenced philosophers. I am thinking mainly of the effective contacts between science and philosophy.

This division of territory between science and philosophy was not a simple business; and in fact it illustrated the weakness of the whole cut-and-dried presupposition upon which it rested. We are aware of nature as an interplay of bodies, colours, sounds, scents, tastes, touches and other various bodily feelings, displayed as in space, in patterns of mutual separation by intervening volumes, and of individual shape. Also the whole is a flux, changing with the lapse of time. This systematic totality is disclosed to us as one complex of things. But the seventeenth-century dualism cuts straight across it. The objective world of science was confined to mere spatial material with simple location in space and time, and subjected to definite rules as to its locomotion. The subjective world of philosophy annexed the colours, sounds, scents, tastes, touches, bodily feelings, as forming the subjective content of the cogitations of the individual minds. Both worlds shared in the general flux; but time, as measured, is assigned by Descartes to the cogitations of the observer's mind. There is obviously one fatal weakness to this scheme. The cogitations of mind exhibit themselves as holding up entities, such as colours for instance, before the mind as the termini of contemplation. But in this theory these colours are, after all, merely the furniture of the mind. Accordingly, the mind seems to be confined to its own private world of cogitations. The subject-object conformation of experience in its entirety lies within the mind as one of its private passions. This conclusion from the Cartesian data is the starting-point from which Berkeley, Hume, and Kant developed their respective systems. And, antecedently to them, it was the point upon which Locke concentrated as being the vital question. Thus the question as to how any knowledge is obtained of the truly objective world of science becomes a problem of the first magnitude. Descartes states that the objective body is perceived by the intellect. He says (second *Meditation*):

> I must, therefore, admit that I cannot even comprehend by imagination what the piece of wax is, and that it is the mind alone which perceives it. I speak of one piece in particular; for, as to wax in general, this is still more evident. But what is the piece of wax that can be perceived only by the mind? . . . The perception of it is neither an act of sight, of touch, nor of imagination, and never was either of these, though it might formerly seem so, but is simply an *intuition* (*inspectio*) of the mind . . .

It must be noted that the Latin word *inspectio* is associated in its classical use with the notion of theory as opposed to practice.

The two great preoccupations of modern philosophy now lie clearly before us. The study of mind divides into psychology, or the study of mental functionings as considered in themselves and in their mutual relations, and into epistemology, or the theory of the knowledge of a common objective world. In other words, there is the study of the cogitations, *qua* passions of the mind, and their study *qua* leading to an inspection (intuition) of an objective world. This is a very uneasy division, giving rise to a host of perplexities whose consideration has occupied the intervening centuries.

As long as men thought in terms of physical notions for the objective world and of mentality for the subjective world, the setting out of the problem, as achieved by Descartes, sufficed as a starting-point. But the balance has been upset by the rise of physiology. In the seventeenth century men passed from the study of physics to the study of philosophy. Towards the end of the nineteenth century, notably in Germany, men passed from the study of physiology to the study of psychology. The change in tone has been decisive. Of course, in the earlier period the intervention of the human body was fully considered, for example, by Descartes in Part V of the *Discourse on Method*. But the physiological instinct had not been developed. In considering the human body, Descartes thought with the outfit of a physicist; whereas the modern psychologists are clothed with the mentalities of medical physiologists. The career of William James is an example of this change in standpoint. He also possessed the clear, incisive genius which could state in a flash the exact point at issue.

The reason why I have put Descartes and James in close juxtaposition is now evident. Neither philosopher finished an epoch by a final solution of a problem. Their great merit is of the opposite sort. They each of them open an epoch by their clear formulation of terms in which thought could profitably express itself at particular stages of knowledge, one for the seventeenth century, the other for the twentieth century. In this respect, they are both to be contrasted with St Thomas Aquinas, who expressed the culmination of Aristotelian scholasticism.

In many ways neither Descartes nor James were the most characteristic philosophers of their respective epochs. I should be disposed to ascribe these positions to Locke and to Bergson respectively, at least so far as concerns their relations to the science of their times. Locke developed the lines of thought which kept philosophy on the move; for example, he emphasized the appeal to psychology. He initiated the age of epoch-making inquiries into urgent problems of limited scope. Undoubtedly, in so doing, he infected philosophy with something of the anti-rationalism of science. But the very groundwork of a fruitful methodology is to start from those clear postulates which must be held to be ultimate so far as concerns the occasion in question. The criticism of such methodological postulates is thus reserved for another opportunity. Locke discovered that the philosophical situation bequeathed by Descartes involved the problems of epistemology and psychology.

Bergson introduced into philosophy the organic conceptions of physiological science. He has most completely moved away from the static materialism of the seventeenth century. His protest against spatialization is a protest against taking the Newtonian conception of nature as being anything except a high abstraction. His so-called anti-intellectualism should be construed in this sense. In some respects he recurs to Descartes; but the recurrence is accompanied with an instinctive grasp of modern biology.

There is another reason for associating Locke and Bergson. The germ of an organic theory of nature is to be found in Locke. His most recent expositor, Professor Gibson,[4] states that Locke's way of conceiving the identity of self-consciousness 'like that of a living organism, involves a genuine transcending of the mechanical view of nature and of mind, embodied in the composition theory'. But it is to be noticed that in the first place Locke wavers in his grasp of this position; and in the second place, what is more important still, he only applies his idea to self-consciousness. The physiological attitude has not yet established itself. The effect of physiology was to put mind back into nature. The neurologist traces first the effect of stimuli

[4] Cf. his book, *Locke's Theory of Knowledge and its Historical Relations*, Camb. Univ. Press, 1917.

along the bodily nerves, then integration at nerve centres, and finally the rise of a projective reference beyond the body with a resulting motor efficacy in renewed nervous excitement. In biochemistry, the delicate adjustment of the chemical composition of the parts to the preservation of the whole organism is detected. Thus the mental cognition is seen as the reflective experience of a totality, reporting for itself what it is in itself as one unit occurrence. This unit is the integration of the sum of its partial happenings, but it is not their numerical aggregate. It has its own unity as an event. This total unity, considered as an entity for its own sake, is the prehension into unity of the patterned aspects of the universe of events. Its knowledge of itself arises from its own relevance to the things of which it prehends the aspects. It knows the world as a system of mutual relevance, and thus sees itself as mirrored in other things. These other things include more especially the various parts of its own body.

It is important to discriminate the bodily pattern, which endures, from the bodily event, which is pervaded by the enduring pattern, and from the parts of the bodily event. The parts of the bodily event are themselves pervaded by their own enduring patterns, which form elements in the bodily pattern. The parts of the body are really portions of the environment of the total bodily event, but so related that their mutual aspects, each in the other, are peculiarly effective in modifying the pattern of either. This arises from the intimate character of the relation of whole to part. Thus the body is a portion of the environment for the part, and the part is a portion of the environment for the body; only they are peculiarly sensitive, each to modifications of the other. This sensitiveness is so arranged that the part adjusts itself to preserve the stability of the pattern of the body. It is a particular example of the favourable environment shielding the organism. The relation of part to whole has the special reciprocity associated with the notion of organism, in which the part is for the whole;

but this relation reigns throughout nature and does not start with the special case of the higher organisms.

Further, viewing the question as a matter of chemistry, there is no need to construe the actions of each molecule in a living body by its exclusive particular reference to the pattern of the complete living organism. It is true that each molecule is affected by the aspect of this pattern as mirrored in it, so as to be otherwise than what it would have been if placed elsewhere. In the same way, under some circumstances an electron may be a sphere, and under other circumstances an egg-shaped volume. The mode of approach to the problem, so far as science is concerned, is merely to ask if molecules exhibit in living bodies properties which are not to be observed amid inorganic surroundings. In the same way, in a magnetic field soft iron exhibits magnetic properties which are in abeyance elsewhere. The prompt self-preservative actions of living bodies, and our experience of the physical actions of our bodies following the determinations of will, suggest the modification of molecules in the body as a result of the total pattern. It seems possible that there may be physical laws expressing the modification of the ultimate basic organisms when they form part of higher organisms with adequate compactness of pattern. It would, however, be entirely in consonance with the empirically observed action of environments, if the direct effects of aspects between the whole body and its parts were negligible. We should expect transmission. In this way the modification of total pattern would transmit itself by means of a series of modifications of a descending series of parts, so that finally the modification of the cell changes its aspect in the molecule, thus effecting a corresponding alteration in the molecule – or in some subtler entity. Thus the question for physiology is the question of the physics of molecules in cells of different characters.

We can now see the relation of psychology to physiology and to physics. The private psychological field is merely the event considered from its own standpoint. The unity of this field is the unity of the event. But it is the event as one entity, and not the event as a sum of parts. The relations of the parts, to each other and to the whole, are their aspects, each in the other. A body for an external observer is the aggregate of the aspects for him of the body as a whole, and also of the body as a sum of parts. For the external observer the aspects of shape and of sense-objects are dominant, at least for cognition. But we must also allow for the possibility that we can detect in ourselves direct aspects of the mentalities of higher organisms. The claim that the cognition of alien mentalities must necessarily be by means of indirect inferences from aspects of shape and of sense-objects is wholly unwarranted by this philosophy of organism. The fundamental principle is that whatever merges into actuality, implants its aspects in every individual event.

Further, even for self-cognition, the aspects of the parts of our own bodies partly take the form of aspects of shape, and of sense-objects. But that part of the bodily event, in respect to which the cognitive mentality is associated, is for itself the unit psychological field. Its ingredients are not referent to the event itself; they are aspects of what lies beyond that event. Thus the self-knowledge inherent in the bodily event is the knowledge of itself as a complex unity, whose ingredients involve all reality beyond itself, restricted under the limitation of its pattern of aspects. Thus we know ourselves as a function of unification of a plurality of things which are other than ourselves. Cognition discloses an event as being an activity, organizing a real togetherness of alien things. But this psychological field does not depend on its cognition; so that this field is still a unit event as abstracted from its self-cognition.

Accordingly, consciousness will be the function of knowing. But what is known is already a prehension of aspects of the one real universe. These aspects are aspects of other events as mutually modifying, each the others. In the pattern of aspects they stand in their pattern of mutual relatedness.

The aboriginal data in terms of which the pattern weaves itself are the aspects of shapes, of sense-objects, and of other eternal objects whose self-identity is not dependent on the flux of things. Wherever such objects have ingression into the general flux, they interpret events, each to the other. They are here in the perceiver; but, as perceived by him, they convey for him something of the total flux which is beyond himself. The subject-object relation takes its origin in the double role of these eternal objects. They are modifications of the subject, but only in their character of conveying aspects of other subjects in the community of the universe. Thus no individual subject can have independent reality, since it is a prehension of limited aspects of subjects other than itself.

The technical phrase 'subject-object' is a bad term for the fundamental situation disclosed in experience. It is really reminiscent of the Aristotelian 'subject-predicate'. It already presupposes the metaphysical doctrine of diverse subjects qualified by their private predicates. This is the doctrine of subjects with private worlds of experience. If this be granted, there is no escape from solipsism. The point is that the phrase 'subject-object' indicates a fundamental entity underlying the objects. Thus the 'objects', as thus conceived, are merely the ghosts of Aristotelian predicates. The primary situation disclosed in cognitive experience is 'ego-object amid objects'. By this I mean that the primary fact is an impartial world transcending the 'here-now' which marks the ego-object, and transcending the 'now' which is the spatial world of simultaneous realization. It is a world also including the actuality of the past, and the limited potentiality of the future, together with the complete world of abstract potentiality, the realm of eternal objects, which transcends, and finds exemplification in and comparison with, the actual course of realization. The ego-object, as consciousness here-now, is conscious of its experient essence as constituted by its internal relatedness to the world of realities, and to the world of ideas. But the ego-object, in being thus constituted, is within the world of realities, and exhibits itself as an organism requiring the ingression of ideas for the purpose of this status among realities. This question of consciousness must be reserved for treatment on another occasion.

The point to be made for the purpose of the present discussion is that a philosophy of nature as organic must start at the opposite end to that requisite for a materialistic philosophy. The materialistic starting-point is from independently existing substances, matter and mind. The matter suffers modifications of its external relations of locomotion, and the mind suffers modifications of its contemplated objects. There are, in this materialistic theory, two sorts of independent substances, each qualified by their appropriate passions. The organic starting-point is from the analysis of process as the realization of events disposed in an interlocked community. The event is the unit of things real. The emergent enduring pattern is the stabilization of the emergent achievement so as to become a fact which retains its identity throughout the process. It will be noted that endurance is not primarily the property of enduring beyond itself, but of enduring within itself. I mean that endurance is the property of finding its pattern reproduced in the temporal parts of the total event. It is in this sense that a total event carries an enduring pattern. There is an intrinsic value identical for the whole and for its succession of parts. Cognition is the emergence, into some measure of individualized reality, of the general substratum of activity, poising before itself possibility, actuality, and purpose.

It is equally possible to arrive at this organic conception of the world if we start from the fundamental notions of modern physics, instead of, as above, from psychology and physiology. In fact by reason of my own studies in mathematics and mathematical physics, I did in fact arrive at my convictions in this way. Mathematical physics presumes in the first place an electromagnetic field of activity pervading space and time. The laws which condition this field are nothing else than the conditions observed by the general activity of the flux of the world, as it individualizes itself in the events. In physics, there is an abstraction. The science ignores what anything is in itself. Its entities are merely considered in respect to their extrinsic reality, that is to say, in respect to their aspects in other things. But the abstraction reaches even further than that; for it is only the aspects in other things, as modifying the spatio-temporal specifications of the life-histories of those other things, which count. The intrinsic reality of the observer comes in : I mean what the observer is for himself is appealed to. For example, the fact that he will see red or blue enters into scientific statements. But the red which the observer sees does not in truth enter into science. What is relevant is merely the bare diversity of the observer's red experiences from all of his other experiences. Accordingly, the intrinsic character of the observer is merely relevant in order to fix the self-identical individuality of the physical entities. These entities are only considered as agencies in fixing the routes in space and in time of the life-histories of enduring entities.

The phraseology of physics is derived from the materialistic ideas of the seventeenth century. But we find that, even in its extreme abstraction, what it is really presupposing is the organic theory of aspects as explained above. First, consider any event in empty space where the word 'empty' means devoid of electrons, or protons, or of any other form of electric charge. Such an event has three roles in physics. In the first place, it is the actual scene of an adventure of energy, either as its habitat or as the locus of a particular stream of energy : anyhow, in this role the energy is there, either as located in space during the time considered, or as streaming through space.

In its second role, the event is a necessary link in the pattern of transmission, by which the character of every event receives some modification from the character of every other event.

In its third role, the event is the repository of a possibility, as to what would happen to an electric charge, either by way of deformation or of locomotion, if it should have happened to be there.

If we modify our assumption by considering an event which includes in itself a portion of the life-history of an electric charge, then the analysis of its three roles still remains; except that the possibility embodied in the third role is now transformed into an actuality. In this replacement of possibility by actuality, we obtain the distinction between empty and occupied events.

Recurring to the empty events, we note the deficiency in them of individuality of intrinsic content. Considering the first role of an empty event, as being a habitat of energy, we note that there is no individual discrimination of an individual bit of energy, either as statically located, or as an element in the stream. There is simply a quantitative determination of activity, without individualization of the activity in itself. This lack of individualization is still more evident in the second and third roles. An empty event is something in itself, but it fails to realize a stable individuality of content. So far as its content is concerned, the empty event is one realized element in a general scheme of organized activity.

Some qualification is required when the empty event is the scene of the transmission of a definite train of recurrent wave-forms. There is now a definite pattern which remains permanent in the event. We find here the first faint trace of enduring individuality. But it is individuality without the faintest capture of originality: for it is merely a permanence arising solely from the implication of the event in a larger scheme of patterning.

Turning now to the examination of an occupied event, the electron has a determinate individuality. It can be traced throughout its life-history through a variety of events. A collection of electrons, together with the analogous atomic charges of positive electricity, forms a body such as we ordinarily perceive. The simplest body of this kind is a molecule, and a set of molecules forms a lump of ordinary matter, such as a chair, or a stone. Thus a charge of electricity is the mark of individuality of content, as additional to the individuality of an event in itself. This individuality of content is the strong point of the materialistic doctrine.

It can, however, be equally well explained on the theory of organism. When we look into the function of the electric charge, we note that its role is to mark the origination of a pattern which is transmitted through space and time. It is the key of some particular pattern. For example, the field of force in any event is to be constructed by attention to the adventures of electrons and protons, and so also are the streams and distributions of energy. Further, the electric waves find their origin in the vibratory adventures of these charges. Thus the transmitted pattern is to be conceived as the flux of aspects throughout space and time derived from the life-history of the atomic charge. The individualization of the charge arises by a conjunction of two characters, in the first place by the continued identity of its mode of functioning as a key for the determination of a diffusion of pattern; and, in the second place, by the unity and continuity of its life-history.

We may conclude, therefore, that the organic theory represents directly what physics actually does assume respecting its ultimate entities. We also notice the complete futility of these entities, if they are conceived as fully concrete individuals. So far as physics is concerned, they are wholly occupied in moving each other about, and they have no reality outside this function. In particular for physics, there is no intrinsic reality.

It is obvious that the basing of philosophy upon the presupposition of organism must be traced back to Leibniz.[5] His monads are for him the ultimately real entities. But he retained the Cartesian substances with their qualifying passions, as also equally expressing for him the final characterization of real things. Accordingly for him there was no concrete reality of internal relations. He had therefore on his hands two distinct points of view. One was that the final real entity is an organizing activity, fusing ingredients into a unity, so that this unity is the reality. The other point of view is that the final real entities are substances supporting qualities. The first point of view depends upon the acceptance of internal relations binding together all reality. The latter is inconsistent with the reality of such relations. To combine these two points of view, his monads were therefore windowless; and their passions merely mirrored the universe by the divine arrangement of a preestablished harmony. This system thus presupposed an aggregate of independent entities. He did not discriminate the event, as the unit of experience, from the enduring organism as its stabilization into importance, and from the cognitive organism as expressing an increased completeness of individualization. Nor did he admit the many termed relations, relating sense-data to various events in diverse ways. These many termed relations are in fact the perspectives which Leibniz does admit, but only on the condition that they are purely qualities of the organizing monads. The difficulty really arises from the unquestioned acceptance of the notion of simple location as fundamental for space and time, and from the acceptance of the notion of independent individual substance as fundamental for a real entity. The only road open to Leibniz was thus the same as that later taken by Berkeley (in a prevalent interpretation of his meaning), namely an appeal to a *Deus ex machina* who was capable of rising superior to the difficulties of metaphysics.

In the same way as Descartes introduced the tradition

[5] Cf. Bertrand Russell, *The Philosophy of Leibniz,* for the suggestion of this line of thought.

of thought which kept subsequent philosophy in some measure of contact with the scientific movement, so Leibniz introduced the alternative tradition that the entities, which are the ultimate actual things, are in some sense procedures of organization. This tradition has been the foundation of the great achievements of German philosophy. Kant reflected the two traditions, one upon the other. Kant was a scientist, but the schools derivative from Kant have had but slight effect on the mentality of the scientific world. It should be the task of the philosophical schools of this century to bring together the two streams into an expression of the world-picture derived from science, and thereby end the divorce of science from the affirmations of our aesthetic and ethical experiences.

CHAPTER X

ABSTRACTION

In the previous chapters I have been examining the reactions of the scientific movement upon the deeper issues which have occupied modern thinkers. No one man, no limited society of men, and no one epoch can think of everything at once. Accordingly for the sake of eliciting the various impacts of science upon thought, the topic has been treated historically. In this retrospect I have kept in mind that the ultimate issue of the whole story is the patent dissolution of the comfortable scheme of scientific materialism which has dominated the three centuries under review. Accordingly various schools of criticism of the dominant opinions have been stressed; and I have endeavoured to outline an alternative cosmological doctrine, which shall be wide enough to include what is fundamental both for science and for its critics. In this alternative scheme, the notion of material, as fundamental, has been replaced by that of organic synthesis. But the approach has always been from the consideration of the actual intricacies of scientific thought, and of the peculiar perplexities which it suggests.

In the present chapter, and in the immediately succeeding chapter, we will forget the peculiar problems of modern science, and will put ourselves at the standpoint of a dispassionate consideration of the nature of things, antecedently to any special investigation into their details. Such a standpoint is termed 'metaphysical'. Accordingly those readers who find metaphysics, even in two slight chapters, irksome, will do well to proceed at once to the chapter on 'Religion and Science', which resumes the topic of the impact of science on modern thought.

These metaphysical chapters are purely descriptive. Their justification is to be sought, (1) in our direct knowledge of the actual occasions which compose our immediate experience, and (2) in their success as forming a basis for harmonizing our systematized accounts of various types of experience, and (3) in their success as providing the concepts in terms of which an epistemology can be framed. By (3) I mean that an account of the general character of what we know must enable us to frame an account of how knowledge is possible as an adjunct within things known.

In any occasion of cognition, that which is known is an actual occasion of experience, as diversified[1] by reference to a realm of entities which transcend that immediate occasion in that they have analogous or different connections with other occasions of experience. For example a definite shade of red may, in the immediate occasion, be implicated with the shape of sphericity in some definite way. But that shade of red, and that spherical shape, exhibit themselves as transcending that occasion, in that either of them has other relationships to other occasions. Also, apart from the actual occurrence of the same things in other occasions, every actual occasion is set within a realm of alternative inter-connected entities. This realm is disclosed by all the untrue propositions which can be predicated significantly of that occasion. It is the realm of alternative suggestions, whose foothold in actuality transcends each actual occasion. The real relevance of untrue propositions for each actual occasion is disclosed by art, romance, and by criticism in reference to ideals. It is the foundation of the metaphysical position which I am maintaining that the understanding of actuality requires a reference to ideality. The two realms are intrinsically inherent in the total metaphysical situation. The truth that some proposition respecting an actual occasion is untrue may express the vital truth as to the aesthetic achievement. It expresses the 'great refusal' which is its primary characteristic. An event is de-

[1] Cf. my *Principles of Natural Knowledge*, Ch. v, Sec. 13.

cisive in proportion to the importance (for it) of its untrue propositions : their relevance to the event cannot be dissociated from what the event is in itself by way of achievement. These transcendent entities have been termed 'universals'. I prefer to use the term 'eternal objects', in order to disengage myself from presuppositions which cling to the former term owing to its prolonged philosophical history. Eternal objects are thus, in their nature, abstract. By 'abstract' I mean that what an eternal object is in itself – that is to say, its essence – is comprehensible without reference to some one particular occasion of experience. To be abstract is to transcend particular concrete occasions of actual happening. But to transcend an actual occasion does not mean being disconnected from it. On the contrary, I hold that each eternal object has its own proper connection with each such occasion, which I term its mode of ingression into that occasion. Thus an eternal object is to be comprehended by acquaintance with (1) its particular individuality, (2) its general relationships to other eternal objects as apt for realization in actual occasions, and (3) the general principle which expresses its ingression in particular actual occasions.

These three headings express two principles. The first principle is that each eternal object is an individual which, in its own peculiar fashion, is what it is. This particular individuality is the individual essence of the object, and cannot be described otherwise than as being itself. Thus the individual essence is merely the essence considered in respect to its uniqueness. Further, the essence of an eternal object is merely the eternal object considered as adding its own unique contribution to each actual occasion. This unique contribution is identical for all such occasions in respect to the fact that the object in all modes of ingression is just its identical self. But it varies from one occasion to another in respect to the differences of its modes of ingression. Thus the metaphysical status of an eternal object is that of a possibility for an actuality. Every actual occasion is defined as to its character by how these possibilities are actualized for that occasion. Thus actualization is a selection among possibilities. More accurately, it is a selection issuing in a gradation of possibilities in respect to their realization in that occasion. This conclusion brings us to the second metaphysical principle: An eternal object, considered as an abstract entity, cannot be divorced from its reference to other eternal objects, and from its reference to actuality generally; though it is disconnected from its actual modes of ingression into definite actual occasions. This principle is expressed by the statement that each eternal object has a 'relational essence'. This relational essence determines how it is possible for the object to have ingression into actual occasions.

In other words: if *A* be an eternal object, then what *A* is in itself involves *A*'s status in the universe, and *A* cannot be divorced from this status. In the essence of *A* there stands a determinateness as to the relationships of *A* to other eternal objects, and an indeterminateness as to the relationships of *A* to actual occasions. Since the relationships of *A* to other eternal objects stand determinately in the essence of *A*, it follows that they are internal relations. I mean by this that these relationships are constitutive of *A*; for an entity which stands in internal relations has no being as an entity not in these relations. In other words, once with internal relations, always with internal relations. The internal relationships of *A* conjointly form its significance.

Again an entity cannot stand in external relations unless in its essence there stands an indeterminateness which is in patience for such external relations. The meaning of the term 'possibility' as applied to *A* is simply that there stands in the essence of *A* a patience for relationships to actual occasions. The relationships of *A* to an actual occasion are simply how the eternal relationships of *A* to other eternal objects are graded as to their realization in that occasion.

Thus the general principle which expresses *A*'s ingression in the particular actual occasion α is the indeterminateness which stands in the essence of *A* as to its ingression into *α*, and is the determinateness which stands in the essence of *α* as to the ingression of *A* into *α*. Thus the synthetic prehension, which is *α*, is the solution of the indeterminateness of *A* into the determinateness of *α*. Accordingly the relationship between *A* and *α* is external as regards *A*, and is internal as regards *α*. Every actual occasion *α* is the solution of all modalities into actual categorical ingressions: truth and falsehood take the place of possibility. The complete ingression of *A* into *α* is expressed by all the true propositions which are about *A* and *α*, and also – it may be – about other things.

The determinate relatedness of the eternal object *A* to every other eternal object is how *A* is systematically and by the necessity of its nature related to every other eternal object. Such relatedness represents a possibility for realization. But a relationship is a fact which concerns all the implicated *relata*, and cannot be isolated as if involving only one of the *relata*. Accordingly there is a general fact of systematic mutual relatedness which is inherent in the character of possibility. The realm of eternal objects is properly described as a 'realm', because each eternal object has its status in this general systematic complex of mutual relatedness.

In respect to the ingression of *A* into an actual occasion *α*, the mutual relationships of A to other eternal objects, as thus graded in realization, require for their expression a reference to the status of *A* and of the other eternal objects in the spatio-temporal relationship. Also this status is not expressible (for this purpose) without a reference to the status of *α* and of other actual occasions in the same spatio-temporal relationship. Accordingly the spatio-temporal relationship, in terms of which the actual course of events is to be expressed, is nothing else than a selective limitation within the general systematic relationships among eternal objects. By 'limitation', as applied to the spatio-temporal continuum, I mean those matter-of-fact determinations –

such as the three dimensions of space, and the four dimensions of the spatio-temporal continuum – which are inherent in the actual course of events, but which present themselves as arbitrary in respect to a more abstract possibility. The consideration of these general limitations at the base of actual things, as distinct from the limitations peculiar to each actual occasion, will be more fully resumed in the chapter on 'God'.

Further, the status of all possibility in reference to actuality requires a reference to this spatio-temporal continuum. In any particular consideration of a possibility we may conceive this continuum to be transcended. But in so far as there is any definite reference to actuality, the definite *how* of transcendence of that spatio-temporal continuum is required. Thus primarily the spatio-temporal continuum is a locus of relational possibility, selected from the more general realm of systematic relationship. This limited locus of relational possibility expresses one limitation of possibility inherent in the general system of the process of realization. Whatever possibility is generally coherent with that system falls within this limitation. Also whatever is abstractedly possible in relation to the general course of events – as distinct from the particular limitations introduced by particular occasions – pervades the spatio-temporal continuum in every alternative spatial situation and at all alternative times.

Fundamentally, the spatio-temporal continuum is the general system of relatedness of all possibilities, in so far as that system is limited by its relevance to the general fact of actuality. Also it is inherent in the nature of possibility that it should include this relevance to actuality. For possibility is that in which there stands achievability, abstracted from achievement.

It has already been emphasized that an actual occasion is to be conceived as a limitation; and that this process of limitation can be still further characterized as a gradation. This characteristic of an actual occasion (α, say) requires further elucidation: An indeterminateness stands in the essence of any eternal object (*A*, say). The actual occasion *α* synthesizes in itself every eternal object; and, in so doing, it includes the *complete* determinate relatedness of *A* to every other eternal object, or set of eternal objects. This synthesis is a limitation of realization but *not* of content. Each relationship preserves its inherent self-identity. But grades of entry into this synthesis are inherent in each actual occasion, such as *α*. These grades can be expressed only as relevance of value. This relevance of value varies – as comparing different occasions – in grade from the inclusion of the individual essence of *A* as an element in the aesthetic synthesis (in some grade of inclusion) to the lowest grade which is the exclusion of the individual essence of *A* as an element in the aesthetic synthesis. In so far as it stands in this lowest grade, every determinate relationship of *A* is merely ingredient in the occasion in respect to the determinate *how* this relationship is an unfulfilled alternative, not contributing any aesthetic value, except as forming an element in the systematic substratum of unfulfilled content. In a higher grade, it may remain unfulfilled, but be aesthetically relevant.

Thus *A*, conceived merely in respect to its relationships to other eternal objects, is '*A* conceived as *not-being*'; where 'not-being' means 'abstracted from the determinate fact of inclusions in, and exclusions from, actual events'. Also '*A* as *not-being* in respect to a definite occasion *α*' means that *A* in all its determinate relationships is excluded from *α*. Again '*A as being* in respect to *α*' means that *A* in some of its determinate relationships is included in *α*. But there can be no occasion which includes *A* in all its determinate relationships; for some of these relationships are contraries. Thus, in regard to excluded relationships, *A* will be *not-being* in *α*, even when in regard to other relationships *A* will be *being* in *α*. In this sense, every occasion is a synthesis of *being* and *not-being*. Furthermore, though some eternal objects are synthesized in an occasion *α* merely *qua not-being*, each eternal object which is synthesized *qua being* is also synthesized *qua not-being*. *'Being'* here means 'individually effective in the aesthetic synthesis'. Also the 'aesthetic synthesis' is the 'experient synthesis' viewed as self-creative, under the limitations laid upon it by its internal relatedness to all other actual occasions. We thus conclude – what has already been stated above – that the general fact of the synthetic prehension of all eternal objects into every occasion wears the double aspect of the indeterminate relatedness of each eternal object to occasions generally, and of its determinate relatedness to each particular occasion. This statement summarizes the account of how external relations are possible. But the account depends upon disengaging the spatio-temporal continuum from its mere implication in actual occasions – according to the usual explanation – and upon exhibiting it in its origin from the general nature of abstract possibility, as limited by the general character of the actual course of events.

The difficulty which arises in respect to internal relations is to explain how any particular truth is possible. In so far as there are internal relations, everything must depend upon everything else. But if this be the case, we cannot know about anything till we equally know everything else. Apparently, therefore, we are under the necessity of saying everything at once. This supposed necessity is palpably untrue. Accordingly it is incumbent on us to explain how there can be internal relations, seeing that we admit finite truths.

Since actual occasions are selections from the realm of possibilities, the ultimate explanation of how actual occasions have the general character which they do have, must lie in an analysis of the general character of the realm of possibility.

The analytical character of the realm of eternal objects is the primary metaphysical truth concerning it. By this character it is meant that the status of any eternal object *A* in this realm is capable of analysis into an indefinite number of subordinate relationships of limited scope. For example if *B* and *C* are two other eternal objects, then there is some perfectly definite relationship *R*(*A*, *B*, *C*) which involves *A*, *B*, *C* only, as to require the mention of no other definite eternal objects in the capacity of *relata.* Of course, the relationship *R*(*A*, *B*, *C*) may involve subordinate relationships which are themselves eternal objects, and *R*(*A*, *B*, *C*) is also itself an eternal object. Also there will be other relationships which in the same sense involve only *A*, *B*, *C*. We have now to examine how, having regard to the internal relatedness of eternal objects, this limited relationship *R*(*A*, *B*, *C*) is possible.

The reason for the existence of finite relationships in the realm of eternal objects is that relationships of these objects among themselves are entirely unselective, and are systematically complete. We are discussing possibility; so that every relationship which is possible is thereby in the realm of possibility. Every such relationship of each eternal object is founded upon the perfectly definite status of that object as a *relatum* in the general scheme of relationships. This definite status is what I have termed the 'relational essence' of the object. This relational essence is determinable by reference to that object alone, and does not require reference to any other objects, except those which are specifically involved in its individual essence when that essence is complex (as will be explained immediately). The meaning of the words 'any' and 'some' springs from this principle—that is to say, the meaning of the 'variable' in logic. The whole principle is that a particular determination can be made of the *how* of some definite relationship of a definite eternal object *A* to a definite finite number *n* of other eternal objects, *without* any determination of the other *n* objects, $X_1, X_2 \ldots X_n$, except that they have, each of them, the requisite status to play their respective parts in that multiple relationship. This principle depends on the fact that the relational essence of an eternal object is not unique to that object. The mere relational essence of each eternal object determines the complete uniform scheme of relational essences, since each object stands internally in all its possible relationships. Thus the realm of possibility provides a uniform scheme of relationships among finite sets of eternal objects; and all eternal objects stand in all such relationships, so far as the status of each permits.

Accordingly the relationships (as in possibility) do not involve the individual essences of the eternal objects; they involve *any* eternal objects as *relata*, subject to the proviso that these *relata* have the requisite relational essences. (It is this proviso which, automatically and by nature of the case, limits the 'any' of the phrase 'any eternal objects'.) This principle is the principle of the isolation of eternal objects in the realm of possibility. The eternal objects are isolated, because their relationships as possibilities are expressible without reference to their respective individual essences. In contrast to the realm of possibility the inclusion of eternal objects within an actual occasion means that in respect to some of their possible relationships there is a togetherness of their individual essences. This realized togetherness is the achievement of an emergent value defined – or, shaped – by the definite eternal relatedness in respect to which the real togetherness is achieved. Thus the eternal relatedness is the form – the ειδος – ; the emergent actual occasion is the *superject* of informed value; value, as abstracted from any particular superject, is the abstract matter – the υλη – which is common to all actual occasions; and the synthetic activity which prehends valueless possibility into superjicient informed value is the substantial activity. This substantial activity is that which is omitted in any analysis of the static factors in the metaphysical situation. The analysed elements of the situation are the attributes of the substantial activity.

The difficulty inherent in the concept of finite internal relations among eternal objects is thus evaded by two metaphysical principles, (1) that the relationships of any eternal object *A*, considered as constitutive of *A*, merely involve other eternal objects as bare *relata* without reference to their individual essences, and (2) that the divisibility of the general relationship of *A* into a multiplicity of finite relationships of *A* stands therefore in the essence of that eternal object. The second principle obviously depends upon the first. To understand *A* is to understand the *how* of a general scheme of relationship. This scheme of relationship does not require the individual uniqueness of the other *relata* for its comprehension. This scheme also discloses itself as being analysable into a multiplicity of limited relationships which have their own individuality and yet at the same time presupposes the total relationship within possibility. In respect to actuality there is first the general limitation of relationships, which reduces this general unlimited scheme to the four dimensional spatio-temporal scheme. This spatio-temporal scheme is, so to speak, the greatest common measure of the schemes of relationship (as limited by actuality) inherent in all the eternal objects. By this it is meant that, *how* select relationships of an eternal object (*A*) are realized in any actual occasion, is always explicable by expressing the status of *A* in respect to this spatio-temporal scheme, and by expressing in this scheme the relationship of the actual occasion to other actual occasions. A definite finite relationship involving the definite eternal objects of a limited set of such objects is itself an eternal object: it is those eternal objects as in that relationship. I will call such an eternal object 'complex'. The eternal objects which are the *relata* in a complex eternal object will be called the 'components' of that eternal object. Also if any of these *relata* are themselves complex, their components will be called 'derivative components' of the original complex object. Also the components of derivative components will also be called derivative components of the original object. Thus the complexity of an eternal object means its analysability into a relationship of component eternal objects. Also the analysis of the general scheme of relatedness of eternal objects means its exhibition as a multiplicity of complex eternal objects. An eternal object, such as a definite shade of green, which cannot be analysed into a relationship of components, will be called 'simple'.

We can now explain how the analytical character of the realm of eternal objects allows of an analysis of that realm into grades.

In the lowest grade of eternal objects are to be placed those objects whose individual essences are simple. This is the grade of zero complexity. Next consider any set of such objects, finite or infinite as to the number of its members. For example, consider the set of three eternal objects, A, B, C, of which none is complex. Let us write $R(A, B, C)$ for some definite possible relatedness of A, B, C. To take a simple example, A, B, C may be three definite colours with the spatio-temporal relatedness to each other of three faces of a regular tetrahedron, anywhere at any time. Then $R(A, B, C)$ is another eternal object of the lowest complex grade. Analogously there are eternal objects of successively higher grades. In respect to any complex eternal object, $S(D_1, D_2, \dots D_n)$, the eternal objects $D_1, \dots D_n$, whose individual essences are constitutive of the individual essence of $S(D_1, \dots D_n)$, are called the components of $S(D_1, \dots D_n)$. It is obvious that the grade of complexity to be ascribed to $S(D_1, \dots D_n)$, is to be taken as one above the highest grade of complexity to be found among its components.

There is thus an analysis of the realm of possibility into simple eternal objects, and into various grades of complex eternal objects. A complex eternal object is an abstract situation. There is a double sense of 'abstraction', in regard to the abstraction of *definite* eternal objects, i.e. non-mathematical abstraction. There is abstraction from actuality, and abstraction from possibility. For example, A and $R(A, B, C)$ are both abstractions from the realm of possibility. Note that A must mean A in all its possible relationships, and among them $R(A, B, C)$. Also $R(A, B, C)$ means $R(A, B, C)$ in all its relationships. But this meaning of $R(A, B, C)$ excludes other relationships into which A can enter. Hence A as in

$R(A, B, C)$ is more abstract than *A simpliciter*. Thus as we pass from the grade of simple eternal objects to higher and higher grades of complexity, we are indulging in higher grades of abstraction from the realm of possibility.

We can now conceive the successive stages of a definite progress towards some assigned mode of abstraction from the realm of possibility, involving a progress (in thought) through successive grades of increasing complexity. I will call any such route of progress 'an abstractive hierarchy'. Any abstractive hierarchy, finite or infinite, is based upon some definite group of simple eternal objects. This group will be called the 'base' of the hierarchy. Thus the base of an abstractive hierarchy is a set of objects of zero complexity. The formal definition of an abstractive hierarchy is as follows:

An 'abstractive hierarchy based upon *g*', where *g* is a group of simple eternal objects, is a set of eternal objects which satisfy the following conditions,

(1) the members of *g* belong to it, and are the only simple eternal objects in the hierarchy,

(2) the components of any complex eternal object in the hierarchy are also members of the hierarchy, and

(3) any set of eternal objects belonging to the hierarchy, whether all of the same grade or whether differing among themselves as to grade, are jointly among the components or derivative components of at least one eternal object which also belongs to the hierarchy.

It is to be noticed that the components of an eternal object are necessarily of a lower grade complexity than itself. Accordingly any member of such a hierarchy, which is of the first grade of complexity, can have as components only members of the group *g*; and any member of the second grade can have as components only members of the first grade, and members of *g*; and so on for the higher grades.

The third condition to be satisfied by an abstractive hierarchy will be called the condition of connexity. Thus an abstractive hierarchy springs from its base; it includes every successive grade from its base either indefinitely onwards, or to its maximum grade; and it is 'connected' by the reappearance (in a higher grade) of any set of its members belonging to lower grades, in the function of a set of components or derivative components of at least one member of the hierarchy.

An abstractive hierarchy is called 'finite' if it stops at a finite grade of complexity. It is called 'infinite' if it includes members belonging respectively to all degrees of complexity.

It is to be noted that the base of an abstractive hierarchy may contain any number of members, finite or infinite. Further, the infinity of the number of the members of the base has nothing to do with the question as to whether the hierarchy be finite or infinite.

A finite abstractive hierarchy will, by definition, possess a grade of maximum complexity. It is characteristic of this grade that a member of it is a component of no other eternal object belonging to any grade of the hierarchy. Also it is evident that this grade of maximum complexity must possess only one member; for otherwise the condition of connexity would not be satisfied. Conversely any complex eternal object defines a finite abstractive hierarchy to be discovered by a process of analysis. This complex eternal object from which we start will be called the 'vertex' of the abstractive hierarchy: it is the sole member of the grade of maximum complexity. In the first stage of the analysis we obtain the components of the vertex. These components may be of varying complexity; but there must be among them at least one member whose complexity is of a grade one lower than that of the vertex. A grade which is one lower than that of a given eternal object will be called the 'proximate grade' for that object. We take then those components of the vertex which belong to its proximate grade; and as the second stage we analyse them into their components. Among these components there must be some belonging to the proximate grade for the objects thus analysed. Add to them the components of the vertex which also belong to this grade of 'second proximation' from the vertex; and, at the third stage analyse as before. We thus find objects belonging to the grade of third proximation from the vertex; and we add to them the components belonging to this grade, which have been left over from the preceding stages of the analysis. We proceed in this way through successive stages, till we reach the grade of simple objects. This grade forms the base of the hierarchy.

It is to be noted that in dealing with hierarchies we are entirely within the realm of possibility. Accordingly the eternal objects are devoid of real togetherness: they remain within their 'isolation'.

The logical instrument which Aristotle used for the analysis of actual fact into more abstract elements was that of classification into species and genera. This instrument has its overwhelmingly important application for science in its preparatory stages. But its use in metaphysical description distorts the true vision of the metaphysical situation. The use of the term 'universal' is intimately connected with this Aristotelian analysis: the term has been broadened of late; but still it suggests that classificatory analysis. For this reason I have avoided it.

In any actual occasion α, there will be a group g of simple eternal objects which are ingredients in that group in the most concrete mode. This complete ingredience in an occasion, so as to yield the most complete fusion of individual essence with other eternal objects in the formation of the individual emergent occasion, is evidently of its own kind and cannot be defined in terms of anything else. But it has a peculiar characteristic which necessarily attaches to it. This characteristic is that there is an *infinite* abstractive hierarchy based upon g which is such that all its members are equally involved in this complete inclusion in α.

The existence of such an infinite abstractive hierarchy is what is meant by the statement that it is impossible to complete the description of an actual occasion by means of concepts. I will call this infinite abstractive hierarchy which is associated with α 'the associated hierarchy of α'. It is also what is meant by the notion of the connectedness of an actual occasion. This connectedness of an occasion is necessary for its synthetic unity and for its intelligibility. There is a connected hierarchy of concepts applicable to the occasion, including concepts of all degrees of complexity. Also in the actual occasion, the individual essences of the eternal objects involved in these complex concepts achieve an aesthetic synthesis, productive of the occasion as an experience for its own sake. This associated hierarchy is the shape, or pattern, or form, of the occasion in so far as the occasion is constituted of what enters into its full realization.

Some confusion of thought has been caused by the fact that abstraction from possibility runs in the opposite direction to an abstraction from actuality, so far as degree of abstractness is concerned. For evidently in describing an actual occasion α, we are nearer to the total concrete fact when we describe α by predicating of it some member of its associated hierarchy, which is of a high grade of complexity. We have then said more about α. Thus, with a high grade of complexity we gain in approach to the full concreteness of α, and with a low grade we lose in this approach. Accordingly the simple eternal objects represent the extreme of abstraction from an actual occasion; whereas simple eternal objects represent the minimum of abstraction from the realm of possibility. It will, I think, be found that, when a high degree of abstraction is spoken of, abstraction from the realm of possibility is what is usually meant – in other words, an elaborate logical construction.

So far I have merely been considering an actual occasion on the side of its full concreteness. It is this side of the occasion in virtue of which it is an event in nature. But a natural event, in this sense of the term, is only an abstraction from a complete actual occasion. A complete occasion includes that which in cognitive experience takes the form of memory, anticipation, imagination, and thought. These elements in an experient occasion are also modes of inclusion of complex eternal objects in the synthetic prehension, as elements in the emergent value. They differ from the concreteness of full inclusion. In a sense this difference is inexplicable; for each mode of inclusion is of its own kind, not to be explained in terms of anything else. But there is a common difference which discriminates these modes of inclusion from the full concrete ingression which has been discussed. This *differentia* is abruptness. By 'abruptness' I mean that what is remembered, or anticipated, or imagined, or thought, is exhausted by a finite complex concept. In each case there is one finite eternal object prehended within the occasion as the vertex of a finite hierarchy. This breaking off from an actual illimitability is what in any occasion marks off that which is termed mental from that which belongs to the physical event to which the mental functioning is referred.

In general there seems to be some loss of vividness in the apprehension of the eternal objects concerned: for example, Hume speaks of 'faint copies'. But this faintness seems to be a very unsafe ground for differentiation. Often things realized in thought are more vivid than the same things in inattentive physical experience. But the things apprehended as mental are always subject to the condition that we come to a stop when we attempt to explore ever higher grades of complexity in their realized relationships. We always find that we have thought of just this – whatever it may be – and of no more. There is a limitation which breaks off the finite concept from the higher grades of illimitable complexity.

Thus an actual occasion is a prehension of one infinite hierarchy (its associated hierarchy) together with various finite hierarchies. The synthesis into the occasion of the infinite hierarchy is according to its specific mode of realization, and that of the finite hierarchies is according to various other specific modes of realization. There is one metaphysical principle which is essential for the rational coherence of this account of the general character of an experient occasion. I call this principle, 'the translucency of realization'. By this I mean that any eternal object is just itself in whatever mode of realization it is involved. There can be no distortion of the individual essence without thereby producing a different eternal object. In the essence of each eternal object there stands an indeterminateness which expresses its indifferent patience for any mode of ingression into any actual occasion. Thus in cognitive experience, there can be the cognition of the same eternal object as in the same occasion having ingression with implication in more than one grade of realization. Thus the translucency of realization, and the possible multiplicity of modes of ingression into the same occasion, together form the foundation for the correspondence theory of truth.

In this account of an actual occasion in terms of its connection, with the realm of eternal objects, we have gone back to the train of thought in our second chapter, where the nature of mathematics was discussed. The idea, ascribed to Pythagoras, has been amplified, and put forward as the first chapter in metaphysics. The next chapter is concerned with the puzzling fact that there is an actual course of events which is in itself a limited fact, in that metaphysically speaking it might have been otherwise. But other metaphysical investigations are omitted; for example, epistemology, and the classification of some elements in the unfathomable wealth of the field of possibility. This last topic brings metaphysics in sight of the special topics of the various sciences.

CHAPTER XI

GOD

Aristotle found it necessary to complete his metaphysics by the introduction of a prime mover – God. This, for two reasons, is an important fact in the history of metaphysics. In the first place if we are to accord to anyone the position of the greatest metaphysician, having regard to genius of insight, to general equipment in knowledge, and to the stimulus of his metaphysical ancestry, we must choose Aristotle. Secondly, in his consideration of this metaphysical question he was entirely dispassionate; and he is the last European metaphysician of first-rate importance for whom this claim can be made. After Aristotle, ethical and religious interests began to influence metaphysical conclusions. The Jews dispersed, first willingly and then forcibly, and the Judaic-Alexandrian school arose. Then Christianity, closely followed by Mahometanism, intervened. The Greek gods who surrounded Aristotle were subordinate metaphysical entities, well within nature. Accordingly on the subject of his prime mover, he would have no motive, except to follow his metaphysical train of thought whithersoever it led him. It did not lead him very far towards the production of a God available for religious purposes. It may be doubted whether any properly general metaphysics can ever, without the illicit introduction of other considerations, get much further than Aristotle. But his conclusion does represent a first step without which no evidence on a narrower experimental basis can be of much avail in shaping the conception. For nothing, within any limited type of experience, can give intelligence to shape our ideas of any entity at the base of all actual things, unless the general character of things requires that there be such an entity.

The phrase, prime mover, warns us that Aristotle's thought was enmeshed in the details of an erroneous physics and an erroneous cosmology. In Aristotle's physics special causes were required to sustain the motions of material things. These could easily be fitted into his system, provided that the general cosmic motions could be sustained. For then, in relation to the general working system, each thing could be provided with its true end. Hence the necessity for a prime mover who sustains the motions of the spheres on which depends the adjustment of things. Today we repudiate the Aristotelian physics and the Aristotelian cosmology, so that the exact form of the above argument manifestly fails. But if our general metaphysics is in any way similar to that outlined in the previous chapter, an analogous metaphysical problem arises which can be solved only in an analogous fashion. In the place of Aristotle's God as prime mover, we require God as the principle of concretion. This position can be substantiated only by the discussion of the general implication of the course of actual occasions – that is to say, of the process of realization.

We conceive actuality as in essential relation to an unfathomable possibility. Eternal objects inform actual occasions with hierarchic patterns, included and excluded in every variety of discrimination. Another view of the same truth is that every actual occasion is a limitation imposed on possibility, and that by virtue of this limitation the particular value of that shaped togetherness of things emerges. In this way we express how a single occasion is to be viewed in terms of possibility, and how possibility is to be viewed in terms of a single actual occasion. But there are no single occasions, in the sense of isolated occasions. Actuality is through and through togetherness – togetherness of otherwise isolated eternal objects, and togetherness of all actual occasions. It is my task in this chapter to describe the unity of actual occasions. The previous chapter centred its interest in the abstract: the present chapter deals with the concrete, i.e. that which has grown together.

Consider an occasion α: – we have to enumerate how other actual occasions are in α, in the sense that their relationships with α are constitutive of the essence of α. What α is in itself, is that it is a unit of realized experience; accordingly we ask how other occasions are in the experience which is α. Also for the present I am excluding cognitive experience. The complete answer to this question is, that the relationships among actual occasions are as unfathomable in their variety of type as are those among eternal objects in the realm of abstraction. But there are fundamental types of such relationships in terms of which the whole complex variety can find its description.

A preliminary for the understanding of these types of entry (of one occasion into the essence of another) is to note that they are involved in the modes of realization of abstractive hierarchies, discussed in the previous chapter. The spatio-temporal relationships, involved in those hierarchies as realized in α, have all a definition in terms of α and of the occasions entrant in α. Thus the entrant occasions lend their aspects to the hierarchies, and thereby convert spatio-temporal modalities into categorical determinations; and the hierarchies lend their forms to the occasions and thereby limit the entrant occasions to being entrant only under those forms. Thus in the same way (as seen in the previous chapter) that every occasion is a synthesis of all eternal objects under the limitation of gradations of actuality, so every occasion is a synthesis of all occasions under the limitation of gradations of types of entry. Each occasion synthesizes the totality of content under its own limitations of mode.

In respect to these types of internal relationship between α and other occasions, these other occasions (as constitutive of α) can be classified in many alternative ways. These are all concerned with different definitions of past, present, and future. It has been usual in philosophy to assume that these various definitions must necessarily be equivalent. The present state of opinion in physical science conclusively shows that this assumption is without metaphysical justification even although any such discrimination may be found to be unnecessary for physical science. This question has already been dealt with in the chapter on relativity. But the physical theory of relativity touches only the fringe of the various theories which are metaphysically tenable. It is important for my argument to insist upon the unbounded freedom within which the actual is a unique categorical determination.

Every actual occasion exhibits itself as a process: it is a becomingness. In so disclosing itself, it places itself as one among a multiplicity of other occasions, without which it could not be itself. It also defines itself as a particular individual achievement, focusing in its limited way an unbounded realm of eternal objects.

Any one occasion α issues from other occasions which collectively form its past. It displays for itself other occasions which collectively form its present. It is in respect to its associated hierarchy, as displayed in this immediate present, that an occasion finds its own originality. It is that display which is its own contribution to the output of actuality. It may be conditioned, and even completely determined by the past from which it issues. But its display in the present under those conditions is what directly emerges from its prehensive activity. The occasion α also holds within itself an indetermination in the form of a future, which has partial determination by reason of its inclusion in α and also has determinate spatio-temporal relatedness to α and to actual occasions of the past from α and of the present for α.

This future is a synthesis in α of eternal objects as not-being and as requiring the passage from α to other individualizations (with determinate spatio-temporal relations to α) in which not-being becomes being.

There is also in α what, in the previous chapter, I have termed the 'abrupt' realization of finite eternal objects. This abrupt realization requires either a reference of the basic objects of the finite hierarchy to determinate occasions other than α (as their situations, in past, present, future); or requires a realization of these eternal objects in determinate relationships, but under the aspect of exemption from inclusion in the spatio-temporal scheme of relatedness between actual occasions. This abrupt synthesis of eternal objects in each occasion is the inclusion in actuality of the analytical character of the realm of eternality. This inclusion has those limited gradations of actuality which characterize every occasion by reason of its essential limitation. It is this realized extension of external relatedness beyond the mutual relatedness of the actual occasions, which prehends into each occasion the full sweep of eternal relatedness. I term this abrupt realization the 'graded envisagement' which each occasion prehends into its synthesis. This graded envisagement is how the actual includes what (in one sense) is not-being as a positive factor in its own achievement. It is the source of error, of truth, of art, of ethics, and of religion. By it, fact is confronted with alternatives.

This general concept, of an event as a process whose outcome is a unit of experience, points to the analysis of an event into (1) substantial activity, (2) conditioned potentialities which are there for synthesis, and (3) the achieved outcome of the synthesis. The unity of all actual occasions forbids the analysis of substantial activities into independent entities. Each individual activity is nothing but the mode in which the general activity is individualized by the imposed conditions. The envisagement which enters into the synthesis is also a character which conditions the synthesizing activity. The general activity is not an entity in the sense in which occasions or eternal objects are entities. It is a general metaphysical character which underlies all occasions, in a particular mode for each occasion. There is nothing with which to compare it: it is Spinoza's one infinite substance. Its attributes are its character of individualization into a multiplicity of modes, and the realm of eternal objects which are variously synthesized in these modes. Thus eternal possibility and modal differentiation into individual multiplicity are the attributes of the one substance. In fact each general element of the metaphysical situation is an attribute of the substantial activity.

Yet another element in the metaphysical situation is disclosed by the consideration that the general attribute of modality is limited. This element must rank as an attribute of the substantial activity. In its nature each mode is limited, so as not to be other modes. But, beyond these limitations of particulars, the general modal individualization is limited in two ways: In the first place it is an actual course of events, which might be otherwise so far as concerns eternal possibility, but *is* that course. This limitation takes three forms, (1) the special logical relations which all events must conform to, (2) the selection of relationships to which the events do conform, and (3) the particularity which infects the course even within those general relationships of logic and causation. Thus this first limitation is a limitation of antecedent selection. So far as the general metaphysical situation is concerned, there might have been an indiscriminate modal pluralism apart from logical or other limitation. But there could not then have been these modes, for each mode represents a synthesis of actualities which are limited to conform to a standard. We here come to the second way of limitation. Restriction is the price of value. There cannot be value without antecedent standards of value, to discriminate the acceptance or rejection of what is before the envisaging mode of activity. Thus there is an antecedent limitation among values, introducing contraries, grades, and oppositions.

According to this argument the fact that there is a process of actual occasions, and the fact that the occasions are the emergence of values which require such limitations, both require that the course of events should have developed amid an antecedent limitation composed of conditions, particularization, and standards of value.

Thus as a further element in the metaphysical situation, there is required a principle of limitation. Some particular *how* is necessary, and some particularization in the *what* of matter of fact is necessary. The only alternative to this admission, is to deny the reality of actual occasions. Their apparent irrational limitation must be taken as a proof of illusion and we must look for reality behind the scene. If we reject this alternative behind the scene, we must provide a ground for limitation which stands among the attributes of the substantial activity. This attribute provides the limitation for which no reason can be given : for all reason flows from it. God is the ultimate limitation, and his existence is the ultimate irrationality. For no reason can be given for just that limitation which it stands in his nature to impose. God is not concrete, but he is the ground for concrete actuality. No reason can be given for the nature of God, because that nature is the ground of rationality.

In this argument the point to notice is, that what is metaphysically indeterminate has nevertheless to be categorically determinate. We have come to the limit of rationality. For there is a categorical limitation which does not spring from any metaphysical reason. There is a metaphysical need for a principle of determination, but there can be no metaphysical reason for what is determined. If there were such a reason, there would be no need for any further principle : for metaphysics would already have provided the determination. The general principle of empiricism depends upon the doctrine that there is a principle of concretion which is not discoverable by abstract reason. What further can be known about God must be sought in the region of particular experiences, and therefore rests on an empirical basis. In respect to the interpretation of these experiences, mankind has differed profoundly. He has been named respectively, Jehovah, Allah, Brahma, Father in Heaven, Order of Heaven, First Cause, Supreme

Being, Chance. Each name corresponds to a system of thought derived from the experiences of those who have used it.

Among medieval and modern philosophers, anxious to establish the religious significance of God, an unfortunate habit has prevailed of paying to him metaphysical compliments. He has been conceived as the foundation of the metaphysical situation with its ultimate activity. If this conception be adhered to, there can be no alternative except to discern in him the origin of all evil as well as of all good. He is then the supreme author of the play, and to him must therefore be ascribed its shortcomings as well as its success. If he be conceived as the supreme ground for limitation, it stands in his very nature to divide the good from the evil, and to establish reason 'within her dominions supreme'.

CHAPTER XII

RELIGION AND SCIENCE

The difficulty in approaching the question of the relations between religion and science is, that its elucidation requires that we have in our minds some clear idea of what we mean by either of the terms, 'religion' and 'science'. Also I wish to speak in the most general way possible, and to keep in the background any comparison of particular creeds, scientific or religious. We have got to understand the type of connection which exists between the two spheres, and then to draw some definite conclusions respecting the existing situation which at present confronts the world.

The conflict between religion and science is what naturally occurs to our minds when we think of this subject. It seems as though, during the last half-century, the results of science and the beliefs of religion had come into a position of frank disagreement, from which there can be no escape, except by abandoning either the clear teaching of science, or the clear teaching of religion. This conclusion has been urged by controversialists on either side. Not by all controversialists, of course, but by those trenchant intellects which every controversy calls out into the open.

The distress of sensitive minds, and the zeal for truth, and the sense of the importance of the issues, must command our sincerest sympathy. When we consider what religion is for mankind, and what science is, it is no exaggeration to say that the future course of history depends upon the decision of this generation as to the relations between them. We have here the two strongest general forces (apart from the mere impulse of the various senses) which influence men, and they seem to be set one against the other – the force of our religious intuitions, and the force of our impulse to accurate observation and logical deduction.

A great English statesman once advised his countrymen to use large-scale maps, as a preservative against alarms, panics, and general misunderstanding of the true relations between nations. In the same way in dealing with the clash between permanent elements of human nature, it is well to map our history on a large scale, and to disengage ourselves from our immediate absorption in the present conflicts. When we do this, we immediately discover two great facts. In the first place, there has always been a conflict between religion and science; and in the second place, both religion and science have always been in a state of continual development. In the early days of Christianity, there was a general belief among Christians that the world was coming to an end in the lifetime of people then living. We can make only indirect inferences as to how far this belief was authoritatively proclaimed; but it is certain that it was widely held, and that it formed an impressive part of the popular religious doctrine. The belief proved itself to be mistaken, and Christian doctrine adjusted itself to the change. Again in the early Church individual theologians very confidently deduced from the Bible opinions concerning the nature of the physical universe. In the year AD 535, a monk named Cosmas[1] wrote a book which he entitled *Christian Topography*. He was a travelled man who had visited India and Ethiopia; and finally he lived in a monastery at Alexandria, which was then a great centre of culture. In his book, basing himself upon the direct meaning of Biblical texts as construed by him in a literal fashion, he denied the existence of the antipodes, and asserted that the world is a flat parallelogram whose length is double its breadth.

In the seventeenth century the doctrine of the motion

[1] Cf. Lecky's *The Rise and Influence of Rationalism in Europe*, Ch. III.

of the earth was condemned by a Catholic tribunal. A hundred years ago the extension of time demanded by geological science distressed religious people, Protestant and Catholic. And today the doctrine of evolution is an equal stumbling-block. These are only a few instances illustrating a general fact.

But all our ideas will be in a wrong perspective if we think that this recurring perplexity was confined to contradictions between religion and science; and that in these controversies religion was always wrong, and that science was always right. The true facts of the case are very much more complex, and refuse to be summarized in these simple terms.

Theology itself exhibits exactly the same character of gradual development, arising from an aspect of conflict between its own proper ideas. This fact is a commonplace to theologians, but is often obscured in the stress of controversy. I do not wish to overstate my case; so I will confine myself to Roman Catholic writers. In the seventeenth century, a learned Jesuit, Father Petavius, showed that the theologians of the first three centuries of Christianity made use of phrases and statements which since the fifth century would be condemned as heretical. Also Cardinal Newman devoted a treatise to the discussion of the development of doctrine. He wrote it before he became a great Roman Catholic ecclesiastic; but throughout his life, it was never retracted and continually reissued.

Science is even more changeable than theology. No man of science could subscribe without qualification to Galileo's beliefs, or to Newton's beliefs, or to all his own scientific beliefs of ten years ago.

In both regions of thought, additions, distinctions, and modifications have been introduced. So that now, even when the same assertion is made today as was made a thousand, or fifteen hundred years ago, it is made subject to limitations or expansions of meaning, which were not contemplated at the earlier epoch. We are told by logicians that a proposition must be either true or false, and that there is no middle term. But in practice, we may know that a proposition expresses an important truth, but that it is subject to limitations and qualifications which at present remain undiscovered. It is a general feature of our knowledge, that we are insistently aware of important truths; and yet that the only formulations of these truths which we are able to make presuppose a general standpoint of conceptions which may have to be modified. I will give you two illustrations, both from science: Galileo said that the earth moves and that the sun is fixed; the Inquisition said that the earth is fixed and the sun moves; and Newtonian astronomers, adopting an absolute theory of space, said that both the sun and the earth move. But now we say that any one of these three statements is equally true, provided that you have fixed your sense of 'rest' and 'motion' in the way required by the statement adopted. At the date of Galileo's controversy with the Inquisition, Galileo's way of stating the facts was, beyond question, the fruitful procedure for the sake of scientific research. But in itself it was not more true than the formulation of the Inquisition. But at that time the modern concepts of relative motion were in nobody's mind; so that the statements were made in ignorance of the qualifications required for their more perfect truth. Yet this question of the motions of the earth and the sun expresses a real fact in the universe; and all sides had got hold of important truths concerning it. But with the knowledge of those times, the truths appeared to be inconsistent.

Again I will give you another example taken from the state of modern physical science. Since the time of Newton and Huyghens in the seventeenth century there have been two theories as to the physical nature of light. Newton's theory was that a beam of light consists of a stream of very minute particles, or corpuscles, and that we have the sensation of light when these corpuscles strike the retinas of our eyes. Huyghens' theory was that light consists of very minute waves of trembling in an all-pervading ether, and that these waves are travelling along a beam of light. The two theories are contradictory. In the eighteenth century Newton's theory was believed, in the nineteenth century Huyghens' theory was believed. Today there is one large group of phenomena which can be explained only on the wave theory, and another large group which can be explained only on the corpuscular theory. Scientists have to leave it at that, and wait for the future, in the hope of attaining some wider vision which reconciles both.

We should apply these same principles to the questions in which there is a variance between science and religion. We would believe nothing in either sphere of thought which does not appear to us to be certified by solid reasons based upon the critical research either of ourselves or of competent authorities. But granting that we have honestly taken this precaution, a clash between the two on points of detail where they overlap should not lead us hastily to abandon doctrines for which we have solid evidence. It may be that we are more interested in one set of doctrines than in the other. But, if we have any sense of perspective and of the history of thought, we shall wait and refrain from mutual anathemas.

We should wait: but we should not wait passively, or in despair. The clash is a sign that there are wider truths and finer perspectives within which a reconciliation of a deeper religion and a more subtle science will be found.

In one sense, therefore, the conflict between science and religion is a slight matter which has been unduly emphasized. A mere logical contradiction cannot in itself point to more than the necessity of some readjustments, possibly of a very minor character on both sides. Remember the widely different aspects of events which are dealt with in science and in religion respectively. Science is concerned with the general conditions which are observed to regulate physical phenomena; whereas religion is wholly wrapped up in the contemplation of moral and aesthetic values. On the one side there is the law of gravitation, and on the other the contemplation of the beauty of holiness. What one side sees, the other misses; and vice versa.

Consider, for example, the lives of John Wesley and of St Francis of Assisi. For physical science you have in these lives merely ordinary examples of the operation of the principles of physiological chemistry, and of the dynamics of nervous reactions: for religion you have lives of the most profound significance in the history of the world. Can you be surprised that, in the absence of a perfect and complete phrasing of the principles of science and of the principles of religion which apply to these specific cases, the accounts of these lives from these divergent standpoints should involve discrepancies? It would be a miracle if it were not so.

It would, however, be missing the point to think that we need not trouble ourselves about the conflict between science and religion. In an intellectual age there can be no active interest which puts aside all hope of a vision of the harmony of truth. To acquiesce in discrepancy is destructive of candour, and of moral cleanliness. It belongs to the self-respect of intellect to pursue every tangle of thought to its final unravelment. If you check that impulse, you will get no religion and no science from an awakened thoughtfulness. The important question is, In what spirit are we going to face the issue? There we come to something absolutely vital.

A clash of doctrines is not a disaster – it is an opportunity. I will explain my meaning by some illustrations from science. The weight of an atom of nitrogen was well known. Also it was an established scientific doctrine that the average weight of such atoms in any considerable mass will be always the same. Two experimenters, the late Lord Rayleigh and the late Sir William Ramsay, found that if they obtained nitrogen by two different methods, each equally effective for that purpose, they always observed a persistent slight difference between the average weights of the atoms in the two cases. Now I ask you, would it have been rational of these men to have despaired because of this conflict between chemical theory and scientific observation? Suppose that for some reason the chemical doctrine had been highly prized throughout some district as the foundation of its social order: would it have been wise, would it have been candid, would it have been moral, to forbid the disclosure of the fact that the experiments produced discordant results? Or, on the other hand, should Sir William Ramsey and Lord Rayleigh have proclaimed that chemical theory was now a detected delusion? We see at once that either of these ways would have been a method of facing the issue in an entirely wrong spirit. What Rayleigh and Ramsay did was this: They at once perceived that they had hit upon a line of investigation which would disclose some subtlety of chemical theory that had hitherto eluded observation. The discrepancy was not a disaster: it was an opportunity to increase the sweep of chemical knowledge. You all know the end of the story: finally argon was discovered, a new chemical element which had lurked undetected, mixed with the nitrogen. But the story has a sequel which forms my second illustration. This discovery drew attention to the importance of observing accurately minute differences in chemical substances as obtained by different methods. Further researches of the most careful accuracy were undertaken. Finally another physicist, F. W. Aston, working in the Cavendish Laboratory at Cambridge, in England, discovered that even the same element might assume two or more distinct forms, termed isotopes, and that the law of the constancy of average atomic weight holds for each of these forms, but as between the different isotopes differs slightly. The research has effected a great stride in the power of chemical theory, far transcending in importance the discovery of argon from which it originated. The moral of these stories lies on the surface, and I will leave to you their application to the case of religion and science.

In formal logic, a contradiction is the signal of a defeat:

but in the evolution of real knowledge it marks the first step in progress towards a victory. This is one great reason for the utmost toleration of variety of opinion. Once and for ever, this duty of toleration has been summed up in the words, 'Let both grow together until the harvest.' The failure of Christians to act up to this precept, of the highest authority, is one of the curiosities of religious history. But we have not yet exhausted the discussion of the moral temper required for the pursuit of truth. There are short cuts leading merely to an illusory success. It is easy enough to find a theory, logically harmonious and with important applications in the region of fact, provided that you are content to disregard half your evidence. Every age produces people with clear logical intellects, and with the most praiseworthy grasp of the importance of some sphere of human experience, who have elaborated, or inherited, a scheme of thought which exactly fits those experiences which claim their interest. Such people are apt resolutely to ignore, or to explain away, all evidence which confuses their scheme with contradictory instances. What they cannot fit in is for them nonsense. An unflinching determination to take the whole evidence into account is the only method of preservation against the fluctuating extremes of fashionable opinion. This advice seems so easy, and is in fact so difficult to follow.

One reason for this difficulty is that we cannot think first and act afterwards. From the moment of birth we are immersed in action, and can only fitfully guide it by taking thought. We have, therefore, in various spheres of experience to adopt those ideas which seem to work within those spheres. It is absolutely necessary to trust to ideas which are generally adequate, even though we know that there are subtleties and distinctions beyond our ken. Also apart from the necessities of action, we cannot even keep before our minds the whole evidence except under the guise of doctrines which are incompletely harmonized. We cannot think in terms of an indefinite multiplicity of detail; our evidence can acquire its proper importance only if it comes before us marshalled by general ideas. These ideas we inherit – they form the tradition of our civilization. Such traditional ideas are never static. They are either fading into meaningless formulae, or are gaining power by the new lights thrown by a more delicate apprehension. They are transformed by the urge of critical reason, by the vivid evidence of emotional experience, and by the cold certainties of scientific perception. One fact is certain, you cannot keep them still. No generation can merely reproduce its ancestors. You may preserve the life in a flux of form, or preserve the form amid an ebb of life. But you cannot permanently enclose the same life in the same mould.

The present state of religion among the European races illustrates the statements which I have been making. The phenomena are mixed. There have been reactions and revivals. But on the whole, during many generations, there has been a gradual decay of religious influence in European civilization. Each revival touches a lower peak than its predecessor, and each period of slackness a lower depth. The average curve marks a steady fall in religious tone. In some countries the interest in religion is higher than in others. But in those countries where the interest is relatively high, it still falls as the generations pass. Religion is tending to degenerate into a decent formula wherewith to embellish a comfortable life. A great historical movement on this scale results from the convergence of many causes. I wish to suggest two of them which lie within the scope of this chapter for consideration.

In the first place for over two centuries religion has been on the defensive, and on a weak defensive. The period has been one of unprecedented intellectual progress. In this way a series of novel situations have been produced for thought. Each such occasion has found the religious thinkers unprepared. Something, which has been proclaimed to be vital, has finally, after struggle, distress, and anathema, been modified and otherwise interpreted. The next generation of religious apologists then congratulates the religious world on the deeper insight which has been gained. The result of the continued repetition of this undignified retreat, during many generations, has at last almost entirely destroyed the intellectual authority of religious thinkers. Consider this contrast: when Darwin or Einstein proclaim theories which modify our ideas, it is a triumph for science. We do not go about saying that there is another defeat for science, because its old ideas have been abandoned. We know that another step of scientific insight has been gained.

Religion will not regain its old power until it can face change in the same spirit as does science. Its principles may be eternal, but the expression of those principles requires continual development. This evolution of religion is in the main a disengagement of its own proper ideas from the adventitious notions which have crept into it by reason of the expression of its own ideas in terms of the imaginative picture of the world entertained in previous ages. Such a release of religion from the bonds of imperfect science is all to the good. It stresses its own genuine message. The great point to be kept in mind is that normally an advance in science will show that statements of various religious beliefs require some sort of modification. It may be that they have to be expanded or explained, or indeed entirely restated. If the religion is a sound expression of truth, this modification will only exhibit more adequately the exact point which is of importance. This process is a gain. In so far, therefore, as any religion has any contact with physical facts, it is to be expected that the point of view of those facts must be continually modified as scientific knowledge advances. In this way, the exact relevance of these facts for religious thought will grow more and more clear. The progress of science must result in the unceasing codification of religious thought, to the great advantage of religion.

The religious controversies of the sixteenth and seventeenth centuries put theologians into a most unfortunate state of mind. They were always attacking and defending. They pictured themselves as the garrison of a fort surrounded by hostile forces. All such pictures express half-truths. That is why they are so popular. But they are dangerous. This particular picture fostered a pugnacious party spirit which really expresses an ultimate lack of faith. They dared not modify, because they shirked the task of disengaging their spiritual message from the associations of a particular imagery.

Let me explain myself by an example. In the early medieval times, Heaven was in the sky, and Hell was underground; volcanoes were the jaws of Hell. I do not assert that these beliefs entered into the official formulations: but they did enter into the popular understanding of the general doctrines of Heaven and Hell. These notions were what everyone thought to be implied by the doctrine of the future state. They entered into the explanations of the influential exponents of Christian belief. For example, they occur in the *Dialogues* of Pope Gregory[2] the Great, a man whose high official position is surpassed only by the magnitude of his services to humanity. I am not saying what we ought to believe about the future state. But whatever be the right doctrine, in this instance the clash between religion and science, which has relegated the earth to the position of a second-rate planet attached to a second-rate sun, has been greatly to the benefit of the spirituality of religion by dispersing these medieval fancies.

Another way of looking at this question of the evolution of religious thought is to note that any verbal form of statement which has been before the world for some time discloses ambiguities; and that often such ambiguities strike at the very heart of the meaning. The effective sense in which a doctrine has been held in the past cannot be determined by the mere logical analysis of verbal statements, made in ignorance of the logical trap. You have to

[2] Cf. Gregorovius' *History of Rome in the Middle Ages*, Book III, Ch. III, Vol. II, English trans.

take into account the whole reaction of human nature to the scheme of thought. This reaction is of a mixed character, including elements of emotion derived from our lower natures. It is here that the impersonal criticism of science and of philosophy comes to the aid of religious evolution. Example after example can be given of this motive force in development. For example, the logical difficulties inherent in the doctrine of the moral cleansing of human nature by the power of religion rent Christianity in the days of Pelagius and Augustine – that is to say, at the beginning of the fifth century. Echoes of that controversy still linger in theology.

So far, my point has been this: that religion is the expression of one type of fundamental experiences of mankind: that religious thought develops into an increasing accuracy of expression, disengaged from adventitious imagery: that the interaction between religion and science is one great factor in promoting this development.

I now come to my second reason for the modern fading of interest in religion. This involves the ultimate question which I stated in my opening sentences. We have to know what we mean by religion. The churches, in their presentation of their answers to this query, have put forward aspects of religion which are expressed in terms either suited to the emotional reactions of bygone times or directed to excite modern emotional interests of non-religious character. What I mean under the first heading is that religious appeal is directed partly to excite that instinctive fear of the wrath of a tyrant which was inbred in the unhappy populations of the arbitrary empires of the ancient world, and in particular to excite that fear of an all-powerful arbitrary tyrant behind the unknown forces of nature. This appeal to the ready instinct of brute fear is losing its force. It lacks any directness of response, because modern science and modern conditions of life have taught us to meet occasions of apprehension by a critical analysis of their causes and conditions. Religion is the reaction of human nature to its search for God. The presentation of God under the aspect of power awakens every modern instinct of critical reaction. This is fatal; for religion collapses unless its main positions command immediacy of assent. In this respect the old phraseology is at variance with the psychology of modern civilizations. This change in psychology is largely due to science, and is one of the chief ways in which the advance of science has weakened the hold of the old religious forms of expression. The non-religious motive which has entered into modern religious thought is the desire for a comfortable organization of modern society. Religion has been presented as valuable for the ordering of life. Its claims have been rested upon its function as a sanction to right conduct. Also the purpose of right conduct quickly degenerates into the forming of pleasing social relations. We have here a subtle degradation of religious ideas, following upon their gradual purification under the influence of keener ethical intuitions. Conduct is a by-product of religion – an inevitable by-product, but not the main point. Every great religious teacher has revolted against the presentation of religion as a mere sanction of rules of conduct. St Paul denounced the Law, and Puritan divines spoke of the filthy rags of righteousness. The insistence upon rules of conduct marks the ebb of religious fervour. Above and beyond all things, the religious life is not a research after comfort. I must now state, in all diffidence, what I conceive to be the essential character of the religious spirit.

Religion is the vision of something which stands beyond, behind, and within, the passing flux of immediate things; something which is real, and yet waiting to be realized; something which is a remote possibility, and yet the greatest of present facts; something that gives meaning to all that passes, and yet eludes apprehension; something whose possession is the final good, and yet is beyond all reach; something which is the ultimate ideal, and the hopeless quest.

The immediate reaction of human nature to the religious vision is worship. Religion has emerged into human experience mixed with the crudest fancies of barbaric imagination. Gradually, slowly, steadily the vision recurs in history under nobler form and with clearer expression. It is the one element in human experience which persistently shows an upward trend. It fades and then recurs. But when it renews its force, it recurs with an added richness and purity of content. The fact of the religious vision, and its history of persistent expansion, is our one ground for optimism. Apart from it, human life is a flash of occasional enjoyments lighting up a mass of pain and misery, a bagatelle of transient experience.

The vision claims nothing but worship; and worship is a surrender to the claim for assimilation, urged with the motive force of mutual love. The vision never overrules. It is always there, and it has the power of love presenting the one purpose whose fulfilment is eternal harmony. Such order as we find in nature is never force – it presents itself as the one harmonious adjustment of complex detail. Evil is the brute motive force of fragmentary purpose, disregarding the eternal vision. Evil is overruling, retarding, hurting. The power of God is the worship he inspires. That religion is strong which in its ritual and its modes of thought evokes an apprehension of the commanding vision. The worship of God is not a rule of safety – it is an adventure of the spirit, a flight after the unattainable. The death of religion comes with the repression of the high hope of adventure.

CHAPTER XIII

REQUISITES FOR SOCIAL PROGRESS

It has been the purpose of these lectures to analyse the reactions of science in forming that background of instinctive ideas which control the activities of successive generations. Such a background takes the form of a certain vague philosophy as to the last word about things, when all is said. The three centuries, which form the epoch of modern science, have revolved round the ideas of God, mind, matter, and also of space and time in their characters of expressing simple location for matter. Philosophy has on the whole emphasized mind, and has thus been out of touch with science during the two latter centuries. But it is creeping back into its old importance owing to the rise of psychology and its alliance with physiology. Also, this rehabilitation of philosophy has been facilitated by the recent breakdown of the seventeenth-century settlement of the principles of physical science. But, until that collapse, science seated itself securely upon the concepts of matter, space, time, and latterly, of energy. Also there were arbitrary laws of nature determining locomotion. They were empirically observed, but for some obscure reason were known to be universal. Anyone who in practice or theory disregarded them was denounced with unsparing vigour. This position on the part of scientists was pure bluff, if one may credit them with believing their own statements. For their current philosophy completely failed to justify the assumption that the immediate knowledge inherent in any present occasion throws any light either on its past, or its future.

I have also sketched an alternative philosophy of science in which organism takes the place of matter. For this purpose, the mind involved in the materialist theory dissolves into a function of organism. The psychological field then exhibits what an event is in itself. Our bodily event is an unusually complex type of organism and consequently includes cognition. Further, space and time, in their most concrete signification, become the locus of events. An organism is the realization of a definite shape of value. The emergence of some actual value depends on limitation which excludes neutralizing cross-lights. Thus an event is a matter of fact which by reason of its limitation is a value for itself; but by reason of its very nature it also requires the whole universe in order to be itself.

Importance depends on endurance. Endurance is the retention through time of an achievement of value. What endures is identity of pattern, self-inherited. Endurance requires the favourable environment. The whole of science revolves round this question of enduring organisms.

The general influence of science at the present moment can be analysed under the headings: general conceptions respecting the universe, technological applications, professionalism in knowledge, influence of biological doctrines on the motives of conduct. I have endeavoured in the preceding lectures to give a glimpse of these points. It lies within the scope of this concluding lecture to consider the reaction of science upon some problems confronting civilized societies.

The general conceptions introduced by science into modern thought cannot be separated from the philosophical situation as expressed by Descartes. I mean the assumption of bodies and minds as independent individual substances, each existing in its own right apart from any necessary reference to each other. Such a conception was very concordant with the individualism which had issued from the moral discipline of the Middle Ages. But, though the easy reception of the idea is thus explained, the derivation in itself rests upon a confusion, very natural but none the less unfortunate. The moral discipline had emphasized the intrinsic value of the individual entity. This emphasis had put the notions of the individual and of its experiences into the foreground of thought. At this point the confusion commences. The emergent individual value of each entity is transformed into the independent substantial existence of each entity, which is a very different notion.

I do not mean to say that Descartes made this logical, or rather illogical transition, in the form of explicit reasoning. Far from it. What he did, was first to concentrate upon his own conscious experiences, as being facts within the independent world of his own mentality. He was led to speculate in this way by the current emphasis upon the individual value of his total self. He implicitly transformed this emergent individual value, inherent in the very fact of his own reality, into a private world of passions, or modes, of independent substance.

Also the independence ascribed to bodily substances carried them away from the realm of values altogether. They degenerated into a mechanism entirely valueless, except as suggestive of an external ingenuity. The Heavens had lost the glory of God. This state of mind is illustrated in the recoil of Protestantism from aesthetic effects dependent upon a material medium. It was taken to lead to an ascription of value to what is in itself valueless. This recoil was already in full strength antecedently to Descartes. Accordingly, the Cartesian scientific doctrine of bits of matter, bare of intrinsic value, was merely a formulation, in explicit terms, of a doctrine which was current before its entrance into scientific thought or Cartesian philosophy. Probably this doctrine was latent in the scholastic philosophy, but it did not lead to its consequences till it met with the mentality of northern Europe in the sixteenth century. But science, as equipped by Descartes, gave stability and intellectual status to a point of view which has had very mixed effects upon the moral presuppositions of modern communities. Its good effects arose from its efficiency as a method for scientific researches within those limited regions which were then best suited for exploration. The result was a general clearing of the European mind away from the stains left upon it by the hysteria of remote barbaric ages. This was all to the good, and was most completely exemplified in the eighteenth century.

But in the nineteenth century, when society was undergoing transformation into the manufacturing system, the bad effects of these doctrines have been very fatal. The doctrine of minds, as independent substances, leads directly not merely to private worlds of experience, but also to private worlds of morals. The moral intuitions can be held to apply only to the strictly private world of psychological experience. Accordingly, self-respect and making the most of your own individual opportunities together constituted the efficient morality of the leaders among the industrialists of that period. The Western world is now suffering from the limited moral outlook of the three previous generations.

Also the assumption of the bare valuelessness of mere matter led to a lack of reverence in the treatment of natural or artistic beauty. Just when the urbanization of the Western world was entering upon its state of rapid development, and when the most delicate, anxious consideration of the aesthetic qualities of the new material environment was requisite, the doctrine of the irrelevance of such ideas was at its height. In the most advanced industrial countries, art was treated as a frivolity. A striking example of this state of mind in the middle of the nineteenth century is to be seen in London where the marvellous beauty of the estuary of the Thames, as it curves through the city, is wantonly defaced by the Charing Cross railway bridge, constructed apart from any reference to aesthetic values.

The two evils are: one, the ignoration of the true relation of each organism to its environment; and the other, the habit of ignoring the intrinsic worth of the environment which must be allowed its weight in any consideration of final ends.

Another great fact confronting the modern world is the discovery of the method of training professionals, who specialize in particular regions of thought and thereby progressively add to the sum of knowledge within their respective limitations of subject. In consequence of the success of this professionalizing of knowledge, there are two points to be kept in mind, which differentiate our present age from the past. In the first place, the rate of progress is such that an individual human being, of ordinary length of life, will be called upon to face novel situations which find no parallel in his past. The fixed person for the fixed duties, who in older societies was such a godsend, in the future will be a public danger. In the second place, the modern professionalism in knowledge works in the opposite direction so far as the intellectual sphere is concerned. The modern chemist is likely to be weak in zoology, weaker still in his general knowledge of the Elizabethan drama, and completely ignorant of the principles of rhythm in English versification. It is probably safe to ignore his knowledge of ancient history. Of course I am speaking of general tendencies; for chemists are no worse than engineers, or mathematicians, or classical scholars. Effective knowledge is professionalized knowledge, supported by a restricted acquaintance with useful subjects subservient to it.

This situation has its dangers. It produces minds in a groove. Each profession makes progress, but it is progress in its own groove. Now to be mentally in a groove is to live in contemplating a given set of abstractions. The groove prevents straying across country, and the abstraction abstracts from something to which no further attention is paid. But there is no groove of abstraction which is adequate for the comprehension of human life. Thus in the modern world, the celibacy of the medieval learned class has been replaced by a celibacy of the intellect which is divorced from the concrete contemplation of the complete facts. Of course, no one is merely a mathematician, or merely a lawyer. People have lives outside their professions or their businesses. But the point is the restraint of serious thought within a groove. The remainder of life is treated superficially, with the imperfect categories of thought derived from one profession.

The dangers arising from this aspect of professionalism are great, particularly in our democratic societies. The directive force of reason is weakened. The leading intellects lack balance. They see this set of circumstances, or that set; but not both sets together. The task of coordination is left to those who lack either the force or the character to succeed in some definite career. In short, the specialized functions of the community are performed better and more progressively, but the generalized direction lacks vision. The progressiveness in detail only adds to the danger produced by the feebleness of co-ordination.

This criticism of modern life applies throughout, in whatever sense you construe the meaning of a community. It holds if you apply it to a nation, a city, a district, an institution, a family, or even to an individual. There is a development of particular abstractions, and a contraction of concrete appreciation. The whole is lost in one of its aspects. It is not necessary for my point that I should maintain that our directive wisdom, either as individuals or as communities, is less now than in the past. Perhaps it has slightly improved. But the novel pace of progress requires a greater force of direction if disasters are to be avoided. The point is that the discoveries of the nineteenth century were in the direction of professionalism, so that we are left with no expansion of wisdom and with greater need of it.

Wisdom is the fruit of a balanced development. It is this balanced growth of individuality which it should be the aim of education to secure. The most useful discoveries for the immediate future would concern the furtherance of this aim without detriment to the necessary intellectual professionalism.

My own criticism of our traditional educational methods

is that they are far too much occupied with intellectual analysis, and with the acquirement of formularized information. What I mean is, that we neglect to strengthen habits of concrete appreciation of the individual facts in their full interplay of emergent values, and that we merely emphasize abstract formulations which ignore this aspect of the interplay of diverse values.

In every country the problem of the balance of the general and specialist education is under consideration. I cannot speak with first-hand knowledge of any country but my own. I know that there, among practical educationalists, there is considerable dissatisfaction with the existing practice. Also, the adaptation of the whole system to the needs of a democratic community is very far from being solved. I do not think that the secret of the solution lies in terms of the antithesis between thoroughness in special knowledge and general knowledge of a slighter character. The make-weight which balances the thoroughness of the specialist intellectual training should be of a radically different kind from purely intellectual analytical knowledge. At present our education combines a thorough study of a few abstractions, with a slighter study of a larger number of abstractions. We are too exclusively bookish in our scholastic routine. The general training should aim at eliciting our concrete apprehensions, and should satisfy the itch of youth to be doing something. There should be some analysis even here, but only just enough to illustrate the ways of thinking in diverse spheres. In the Garden of Eden Adam saw the animals before he named them: in the traditional system, children named the animals before they saw them.

There is no easy single solution of the practical difficulties of education. We can, however, guide ourselves by a certain simplicity in its general theory. The student should concentrate within a limited field. Such concentration should include all practical and intellectual acquirements requisite for that concentration. This is the ordinary procedure; and, in respect to it, I should be inclined even to increase the facilities for concentration rather than to diminish them. With the concentration there are associated certain subsidiary studies, such as languages for science. Such a scheme of professional training should be directed to a clear end congenial to the student. It is not necessary to elaborate the qualifications of these statements. Such a training must, of course, have the width requisite for its end. But its design should not be complicated by the consideration of other ends. This professional training can only touch one side of education. Its centre of gravity lies in the intellect, and its chief tool is the printed book. The centre of gravity of the other side of training should lie in intuition without an analytical divorce from the total environment. Its object is immediate apprehension with the minimum of eviscerating analysis. The type of generality, which above all is wanted, is the appreciation of variety of value. I mean an aesthetic growth. There is something between the gross specialized values of the mere practical man, and the thin specialized values of the mere scholar. Both types have missed something; and if you add together the two sets of values, you do not obtain the missing elements. What is wanted is an appreciation of the infinite variety of vivid values achieved by an organism in its proper environment. When you understand all about the sun and all about the atmosphere and all about the rotation of the earth, you may still miss the radiance of the sunset. There is no substitute for the direct perception of the concrete achievement of a thing in its actuality. We want concrete fact with a high light thrown on what is relevant to its preciousness.

What I mean is art and aesthetic education. It is, however, art in such a general sense of the term that I hardly like to call it by that name. Art is a special example. What we want is to draw out habits of aesthetic apprehension. According to the metaphysical doctrine which I have been developing, to do so is to increase the depth of individuality.

The analysis of reality indicates the two factors, activity emerging into individualized aesthetic value. Also the emergent value is the measure of the individualization of the activity. We must foster the creative initiative towards the maintenance of objective values. You will not obtain the apprehension without the initiative, or the initiative without the apprehension. As soon as you get towards the concrete, you cannot exclude action. Sensitiveness without impulse spells decadence, and impulse without sensitiveness spells brutality. I am using the word 'sensitiveness' in its most general signification, so as to include apprehension of what lies beyond oneself; that is to say, sensitiveness to all the facts of the case. Thus 'art' in the general sense which I require is any selection by which the concrete facts are so arranged as to elicit attention to particular values which are realizable by them. For example, the mere disposing of the human body and the eyesight so as to get a good view of a sunset is a simple form of artistic selection. The habit of art is the habit of enjoying vivid values.

But, in this sense, art concerns more than sunsets. A factory, with its machinery, its community of operatives, its social service to the general population, its dependence upon organizing and designing genius, its potentialities as a source of wealth to the holders of its stock is an organism exhibiting a variety of vivid values. What we want to train is the habit of apprehending such an organism in its completeness. It is very arguable that the science of political economy, as studied in its first period after the death of Adam Smith (1790), did more harm than good. It destroyed many economic fallacies, and taught how to think about the economic revolution then in progress. But it riveted on men a certain set of abstractions which were disastrous in their influence on modern mentality. It de-humanized industry. This is only one example of a general danger inherent in modern science. Its methodological procedure is exclusive and intolerant, and rightly so. It fixes attention on a definite group of abstractions, neglects everything else, and elicits every scrap of information and theory which is relevant to what it has retained. This method is triumphant, provided that the abstractions are judicious. But, however triumphant, the triumph is within limits. The neglect of these limits leads to disastrous oversights. The anti-rationalism of science is partly justified, as a preservation of its useful methodology; it is partly mere irrational prejudice. Modern professionalism is the training of minds to conform to the methodology. The historical revolt of the seventeenth century, and the earlier reaction towards naturalism, were examples of transcending the abstractions which fascinated educated society in the Middle Ages. These early ages had an ideal of rationalism, but they failed in its pursuit. For they neglected to note that the methodology of reasoning requires the limitations involved in the abstract. Accordingly, the true rationalism must always transcend itself by recurrence to the concrete in search of inspiration. A self-satisfied rationalism is in effect a form of anti-rationalism. It means an arbitrary halt at a particular set of abstractions. This was the case with science.

There are two principles inherent in the very nature of things, recurring in some particular embodiments whatever field we explore – the spirit of change, and the spirit of conservation. There can be nothing real without both. Mere change without conservation is a passage from nothing to nothing. Its final integration yields mere transient non-entity. Mere conservation without change cannot conserve. For after all, there is a flux of circumstance, and the freshness of being evaporates under mere repetition. The character of existent reality is composed of organisms enduring through the flux of things. The low type of organisms have achieved a self-identity dominating their whole physical life. Electrons, molecules, crystals, belong to this type. They exhibit a massive and complete sameness. In the higher types, where life appears, there is greater complexity. Thus, though there is a complex, enduring pattern, it has retreated into deeper recesses of the total fact. In a sense, the self-identity of a human being is more abstract than that of a crystal. It is the life of the spirit. It relates rather to the individualization of the creative activity; so that the changing circumstances received from the environment, are differentiated from the living personality, and are thought of as forming its perceived field. In truth, the field of perception and the perceiving mind are abstractions which, in the concrete, combine into the successive bodily events. The psychological field, as restricted to sense-objects and passing emotions, is the minor permanence, barely rescued from the non-entity of mere change; and the mind is the major permanence, permeating that complete field, whose endurance is the living soul. But the soul would wither without fertilization from its transient experiences. The secret of the higher organisms lies in their two grades of permanences. By this means the freshness of the environment is absorbed into the permanence of the soul. The changing environment is no longer, by reason of its variety, an enemy to the endurance of the organism. The pattern of the higher organism has retreated into the recesses of the individualized activity. It has become a uniform way of dealing with circumstances; and this way is only strengthened by having a proper variety of circumstances to deal with.

This fertilization of the soul is the reason for the necessity of art. A static value, however serious and important, becomes unendurable by its appalling monotony of endurance. The soul cries aloud for release into change. It suffers the agonies of claustrophobia. The transitions of humour, wit, irreverence, play, sleep, and – above all – of art are necessary for it. Great art is the arrangement of the environment so as to provide for the soul vivid, but transient, values. Human beings require something which absorbs them for a time, something out of the routine which they can stare at. But you cannot subdivide life, except in the abstract analysis of thought. Accordingly, the great art is more than a transient refreshment. It is something which adds to the permanent richness of the soul's self-attainment. It justifies itself both by its immediate enjoyment, and also by its discipline of the inmost being. Its discipline is not distinct from enjoyment, but by reason of it. It transforms the soul into the permanent realization of values extending beyond its former self. This element of transition in art is shown by the restlessness exhibited in its history. An epoch gets saturated by the masterpieces of any one style. Something new must be discovered. The human being wanders on. Yet there is a balance in things. Mere change before the attainment of adequacy of achievement, either in quality or output, is destructive of greatness. But the importance of a living art, which moves on and yet leaves its permanent mark, can hardly be exaggerated.

In regard to the aesthetic needs of civilized society the reactions of science have so far been unfortunate. Its materialistic basis has directed attention to things as opposed to values. The antithesis is a false one, if taken in a concrete sense. But it is valid at the abstract level of ordinary thought. This misplaced emphasis coalesced with the abstractions of political economy, which are in fact the abstractions in terms of which commercial affairs are carried on. Thus all thought concerned with social organization expressed itself in terms of material things and of capital. Ultimate values were excluded. They were politely bowed to, and then handed over to the clergy to be kept for Sundays. A creed of competitive business morality was evolved, in some respects curiously high; but entirely devoid of consideration for the value of human life. The workmen were conceived as mere hands, drawn from the pool of labour. To God's question, men gave the answer of Cain – 'Am I my brother's keeper?'; and they incurred Cain's guilt. This was the atmosphere in which the industrial revolution was accomplished in England, and to a large extent elsewhere. The internal history of England during the last half century has been an endeavour slowly and painfully to undo the evils wrought in the first stage of the new epoch. It may be that civilization will never recover from the bad climate which enveloped the introduction of machinery. This climate pervaded the whole commercial system of the progressive Northern European races. It was partly the result of aesthetic errors of Protestantism and partly the result of scientific materialism, and partly the result of the natural greed of mankind, and partly the result of the abstractions of political economy. An illustration of my point is to be found in Macaulay's Essay criticizing Southey's *Colloquies on Society*. It was written in 1830. Now Macaulay was a very favourable example of men living at that date, or at any date. He had genius; he was kindhearted, honourable, and a reformer. This is the extract:

> We are told, that our age has invented atrocities beyond the imagination of our fathers; that society has been brought into a state compared with which extermination would be a blessing; and all because the dwellings of cotton-spinners are naked and rectangular. Mr Southey has found a way, he tells us, in which the effects of manufactures and agriculture may be compared. And what is this way? To stand on a hill, to look at a cottage and a factory, and to see which is the prettier.

Southey seems to have said many silly things in his book; but, so far as this extract is concerned, he could make a good case for himself if he returned to earth after the lapse of nearly a century. The evils of the early industrial system are now a commonplace of knowledge. The point which I am insisting on is the stone-blind eye with which even the best men of that time regarded the importance of aesthetics in a nation's life. I do not believe that we have as yet nearly achieved the right estimate. A contributory cause, of substantial efficacy to produce this disastrous error, was the scientific creed that matter in motion is the one concrete reality in nature; so that aesthetic values form an adventitious, irrelevant addition.

There is another side to this picture of the possibilities of decadence. At the present moment a discussion is raging as to the future of civilization in the novel circumstances of rapid scientific and technological advance. The evils of the future have been diagnosed in various ways, the loss of religious faith, the malignant use of material power, the degradation attending a differential birth-rate favouring the lower types of humanity, the suppression of aesthetic creativeness. Without doubt, these are all evils, dangerous and threatening. But they are not new. From the dawn of history, mankind has always been losing its religious faith, has always suffered from the malignant use of material power, has always suffered from the infertility of its best intellectual types, has always witnessed the periodical decadence of art. In the reign of the Egyptian king, Tutankhamen, there was raging a desperate religious struggle between Modernists and Fundamentalists; the cave pictures exhibit a phase of delicate aesthetic achievement as superseded by a period of comparative vulgarity; the religious leaders, the great thinkers, the great poets and authors, the whole clerical caste in the Middle Ages, have been notably infertile; finally, if we attend to what actually has happened in the past, and disregard romantic visions of democracies, aristocracies, kings, generals, armies, and merchants, material power has generally been wielded with blindness, obstinacy and selfishness, often with brutal malignancy. And yet, mankind has progressed. Even if you take a tiny oasis of peculiar excellence, the type of modern man who would have most chance of happiness in ancient Greece at its best period is probably (as now) an average professional heavyweight boxer, and not an average Greek scholar from Oxford or Germany. Indeed, the main use of the Oxford scholar would have been his capability of writing an ode in glorification of the boxer. Nothing does more harm in unnerving men for their duties in the present, than the attention devoted to the points of excellence in the past as compared with the average failure of the present day.

But, after all, there have been real periods of decadence; and at the present time, as at other epochs, society is decaying, and there is need for preservative action. Professionals are not new to the world. But in the past, professionals have formed unprogressive castes. The point is that professionalism has now been mated with progress. The world is now faced with a self-evolving system, which it cannot stop. There are dangers and advantages in this situation. It is obvious that the gain in material power affords opportunity for social betterment. If mankind can rise to the occasion, there lies in front a golden age of beneficent creativeness. But material power in itself is ethically neutral. It can equally well work in the wrong direction. The problem is not how to produce great men, but how to produce great societies. The great society will put up the men for the occasions. The materialistic philosophy emphasized the given quantity of material, and thence derivatively the given nature of the environment. It thus operated most unfortunately upon the social conscience of mankind. For it directed almost exclusive attention to the aspect of struggle for existence in a fixed environment. To a large extent the environment is fixed, and to this extent there is a struggle for existence. It is folly to look at the universe through rose-tinted spectacles. We must admit the struggle. The question is, Who is to be eliminated? In so far as we are educators, we have to have clear ideas upon that point; for it settles the type to be produced and the practical ethics to be inculcated.

But during the last three generations, the exclusive direction of attention to this aspect of things has been a disaster of the first magnitude. The watchwords of the nineteenth century have been, struggle for existence, competition, class warfare, commercial antagonism between nations, military warfare. The struggle for existence has been construed into the gospel of hate. The full conclusion to be drawn from a philosophy of evolution is fortunately of a more balanced character. Successful organisms modify their environment. Those organisms are successful which modify their environments so as to assist each other. This law is exemplified in nature on a vast scale. For example, the North American Indians accepted their environment, with the result that a scanty population barely succeeded in maintaining themselves over the whole continent. The European races when they arrived in the same continent pursued an opposite policy. They at once co-operated in modifying their environment. The result is that a population more than twenty times that of the Indian population now occupies the same territory, and the continent is not yet full. Again, there are associations of different species which mutually co-operate. This differentiation of species is exhibited in the simplest physical entities, such as the association between electrons and positive nuclei, and in the whole realm of animate nature. The trees in a Brazilian forest depend upon the association of various species of organisms, each of which is mutually dependent on the other species. A single tree by itself is dependent upon all the adverse chances of shifting circumstances. The wind stunts it: the variations in temperature check its foliage: the rains denude its soil: its leaves are blown away and are lost for the purpose of fertilization. You may obtain individual specimens of fine trees either in exceptional circumstances, or where human cultivation has intervened. But in nature the normal way in which trees flourish is by their association in a forest. Each tree may lose something of its individual perfection of growth, but they mutually assist each other in preserving the conditions for survival. The soil is preserved and shaded; and the microbes necessary for its fertility are neither scorched, nor frozen, nor washed away. A forest is the triumph of the organization of mutually dependent species. Further a species of microbes which kills the forest, also exterminates itself. Again the two sexes exhibit the same advantage of differentiation. In the history of the world, the prize has not gone to those species which specialized in methods of violence, or even in defensive armour. In fact, nature began with producing animals encased in hard shells for defence against the ills of life. It also experimented in size. But smaller animals, without external armour, warm-blooded, sensitive, and alert, have cleared these monsters off the face of the earth. Also, the lions and tigers are not the successful species. There is something in the ready use of force which defeats its own object. Its main defect is that it bars co-operation. Every organism requires an environment of friends, partly to shield it from violent changes, and partly to supply it with its wants. The gospel of force is incompatible with a social life. By force, I mean antagonism in its most general sense.

Almost equally dangerous is the gospel of uniformity. The differences between the nations and races of mankind are required to preserve the conditions under which higher development is possible. One main factor in the upward trend of animal life has been the power of wandering. Perhaps this is why the armour-plated monsters fared badly. They could not wander. Animals wander into new conditions. They have to adapt themselves or die. Mankind has wandered from the trees to the plains, from the plains to the seacoast, from climate to climate, from continent to continent, and from habit of life to habit of life. When man ceases to wander, he will cease to ascend in the scale of being. Physical wandering is still important, but greater still is the power of man's spiritual adventures – adventures of thought, adventures of passionate feeling, adventures of aesthetic experience. A diversification among human communities is essential for the provision of the incentive and material for the Odyssey of the human spirit. Other nations of different habits are not enemies: they are godsends. Men require of their neighbours something sufficiently akin to be understood, something sufficiently different to provoke attention, and something great enough to command admiration. We must not expect, however, all the virtues. We should even be satisfied if there is something odd enough to be interesting.

Modern science has imposed on humanity the necessity for wandering. Its progressive thought and its progressive technology make the transition through time, from generation to generation, a true migration into uncharted seas of adventure. The very benefit of wandering is that it is dangerous and needs skill to avert evils. We must expect, therefore, that the future will disclose dangers. It is the business of the future to be dangerous; and it is among the merits of science that it equips the future for its duties. The prosperous middle classes, who ruled the nineteenth century, placed an excessive value upon placidity of existence. They refused to face the necessities for social reform imposed by the new industrial system, and they are now refusing to face the necessities for intellectual reform imposed by the new knowledge. The middle class pessimism over the future of the world comes from a confusion between civilization and security. In the immediate future there will be less security than in the immediate past, less stability. It must be admitted that there is a degree of instability which is inconsistent with civilization. But, on the whole, the great ages have been unstable ages.

I have endeavoured in these lectures to give a record of a great adventure in the region of thought. It was shared in by all the races of Western Europe. It developed with the slowness of a mass movement. Half a century is its unit of time. The tale is the epic of an episode in the manifestation of reason. It tells how a particular direction of reason emerges in a race by the long preparation of antecedent epochs, how after its birth its subject-matter gradually unfolds itself, how it attains its triumphs, how its influence moulds the very springs of action of mankind, and finally how at its moment of supreme success its limitations disclose themselves and call for a renewed exercise of the creative imagination. The moral of the tale is the power of reason, its decisive influence on the life of humanity. The great conquerors, from Alexander to Caesar, and from Caesar to Napoleon, influenced profoundly the lives of subsequent generations. But the total effect of this influence shrinks to insignificance, if compared to the entire transformation of human habits and human mentality produced by the long line of men of thought from Thales to the present day, men individually powerless, but ultimately the rulers of the world.

INDEX

Fontana Philosophy Classics

This series of texts and anthologies, with substantial introductions, was originated by G. J. Warnock and is being continued under the editorship of A. M. Quinton.

The Principles of Human Knowledge and Three Dialogues Between Hylas and Philonous
George Berkeley *Edited by* G. J. Warnock

An Essay Concerning Human Understanding
John Locke *Edited by* A. D. Woozley

Leviathan
Thomas Hobbes *Edited by* John Plamenatz

A Treatise of Human Nature
David Hume **Book I** *Edited by* D. G. C. Macnabb
Books II and III *Edited by* P. S. Árdal

Hume on Religion
Edited by Richard Wollheim

Russell's Logical Atomism
Edited by David Pears

The Light of Reason
Rationalist Philosophers of the 17th Century
Edited by Martin Hollis

A Guide to the British Moralists
Edited by D. H. Monro

Utilitarianism
Edited by Mary Warnock

A Fontana Selection

The Scientific Renaissance 1450-1630

Marie Boas

This, the first volume in The Rise of Modern Science series, describes the early stages of the Scientific Revolution, beginning with what is traditionally known as the Renaissance of Learning in the fifteenth century. The Scientific Revolution was the effect of a unique series of innovations in scientific ideas and methods; it gave the key to the understanding of the structure and relations of things. It was (and still remains) the greatest intellectual achievement of man since the first stirrings of abstract thought, in that it opened the whole physical universe – an ultimately human nature and behavior – to cumulative exploration.

'Dr. Boas has succeeded brilliantly in getting across one of the most interesting periods of scientific thought. Her book is never dull, and never becomes a mere recitation of facts and information. It is stimulating and even absorbing, and covers a wide field clearly and well.'

The Economist

'Dr. Boas not only lucidly describes the actual achievement of men like Copernicus, Kepler, Galileo and Harvey; she sympathetically shows the psychological repercussions of science upon the whole life and temper of the time.'

Birmingham Post